MEDICINAL PLANTS AND TRADITIONAL MEDICINE IN SIERRA LEONE

MEDICINAL PLANTS AND TRADITIONAL MEDICINE IN SIERRA LEONE

CYRUS MACFOY, PH.D.

Medicinal Plants and Traditional Medicine in Sierra Leone

iUniverse books may be ordered through booksellers or by contacting:

iUniverse
1663 Liberty Drive
Bloomington, IN 47403
www.iuniverse.com
1-800-Authors (1-800-288-4677)

ISBN: 978-1-4917-0609-1 (sc)
ISBN: 978-1-4917-0610-7 (hc)
ISBN: 978-1-4917-0611-4 (e)

Library of Congress Control Number: 2013916234

Print information available on the last page.

iUniverse rev. date: 11/16/2016

CONTENTS

FOREWORD

Medicinal Plants and Traditional Medicine in Sierra Leone is a well written, comprehensive book by an internationally respected Sierra Leonean–American scientist.

I first met author Dr Cyrus MacFoy in 1985 when he joined us at the USDA research center in Beltsville, Maryland as a visiting Professor from the University of Sierra Leone with a United Nations International Atomic Energy Agency fellowship award. Although he was attached to another department (I was leading the Economic Botany Laboratory), he visited me on many occasions for mutual exchanges of information relating to his and my shared interest in Medicinal Plants of the world in general and his in depth knowledge of Sierra Leone. In sharing my knowledge of medicinal plants with him, I was pleased to learn that he had assembled in a book, his three plus decades' research experience on Medicinal Plants and Traditional Medicine in Sierra Leone.

It is a pleasure for me to write this foreword. Cyrus' long overdue book will be an invaluable contribution to the literature It recounts the traditional practices of the peoples of Sierra Leone, and carefully describes the local folklore and traditional medicine. It catalogues more than 200 plant species and more than 100 illnesses and diseases, together with local names, English common names, beliefs, medicinal uses and preparation.. Furthermore it addresses conservation, and how to integrate traditional medicine into the health care system in Sierra Leone. It also addresses some aspects of the chemistry and biological testing of medicinal plants.

Written by a Sierra Leonean from fieldwork conducted throughout the whole of Sierra Leone, Cyrus' book is so readable that non biologists will also find it enjoyable. It will be the most recent addition to the literature from Africa, covering a wide variety of issues, and welcome indeed. This book is indeed informative and timely, in an era when people the world over are turning to natural cures for their various ailments. Coincidentally the search for new drugs from natural sources is also gathering momentum once again. As it well should! Synthetic FDA-approved pharmaceutical drugs here in the US kill more than 100,000 Americans, in hospitals, taken as prescribed. Far fewer humans, if any, are killed annually by herbs in North America, except in rare cases of abuse or misidentity. The phrase "Newly Approved by the FDA" is not a stamp of approval in my view, it is a warning. Do not take for ten years!. More than half of newly FDA-approved drugs are recalled completely or relabelled when they are proved NOT to be safe and efficacious. In other words, long surviving folk practices are safer than newly FDA-approved pharmaceuticals. At least that is my devout belief.

In Africa and other developing countries, large numbers of people employ medicinal plants and traditional medicine already. I often hear and use the figure, 80% of the world's population, by choice or necessity, use traditional healers and medicine. Regrettably some folk practices are not standardized, and may prove unsafe, with little consideration for hygiene. As probably in all endeavors in possibly all countries, there are some profiteering charlatans. Cyrus urges some way to incorporate good traditional medicine into the western allopathic medical system. He emphasizes the urgent need for more research into safety and dosage, for example.

I highly recommend this book to all those who are interested in increasing their knowledge of African traditions in general and Sierra Leonean traditions and culture in particular, its medicinal plants (from a flora similar to those in other west African countries); and those with an interest in continuing the research with a view to developing new drugs from plants. It will prove useful to Sierra Leoneans and other Africans, as well as those interested in African culture and traditions. This American ethnobotanist, Jim Duke, will enjoy seeking out activities and indications shared by those species common to Sierra Leone and the USA, and the frequent chemical rationales for these similarities of activities and indications. And were I lucky enough to travel to Sierra

Leone, I would certainly want a pocket-sized edition with me. Wherever I have traveled, I find it rewarding to relate the local common names with the scientific names. That's when the dialog gets interesting to an ethnobotanist.

Congrats Cyrus, your book has been a long time in the making. Thank goodness it is soon coming to fruition . . . Jim Duke, Oct 10, 2013
James A. Duke, PhD, Ethnobotanist, Green Farmacy Garden, Fulton Md.

ACKNOWLEDGMENTS

I would like to thank Dr. James Duke for writing the foreword and for his encouragement and interest in this book. Many people have contributed to the outcome of this publication, all of whom I am sincerely grateful to. Thus this book should not be regarded as one man's work but rather as a collection of knowledge of the peoples of Sierra Leone. We the scientists should simply regard ourselves as editors who have gathered the information from villagers and put it in readable form.

First and foremost, I would like to thank all the herbalists and villagers who have supplied useful information from their large store of unwritten knowledge; thanks to several of my former students at Fourah Bay College, University of Sierra Leone, including Modupeh Cole, Albert Sama, Edith Cline, Ophilia Barber, Munda Lebbie, and Eric Sumana, for conducting some of the primary research in Sierra Leone. I would also like to thank Michelle Nurse, my former student at the University of Maryland, for acting as my research assistant while putting this material together; Professor N.H.A. Cole for useful discussions, the Smithsonian Institution and the USDA for using some of their graphics and the iUniverse Editorial services for editing this manuscript. The line drawing of *Moringa oleifera* was obtained from *Ethnic Culinary Herbs: A Guide to Identification and Cultivation in Hawaii,* by George W. Staples and Michael S. Kristiansen (University of Hawaii Press, 1999). Finally, I am indebted to my wife, Denise, and children, Namdi and Siatta, for their patience and understanding over the years.

INTRODUCTION

According to the World Health Organization (WHO), traditional medicine (TM) has a central role to play in the twenty-first century. Partnerships between traditional medicine, public health, and health research have great potential, particularly in areas of prevention and management for diseases like HIV-AIDS, TB, malaria, and others, as well as chronic diseases. In 2002, the World Health Organization launched its first ever comprehensive traditional medicine strategy. This strategy was designed to help countries develop their use of TM and complementary and alternate medicine (CAM).

The strategy is designed to assist countries to

- develop national policies on the evaluation and regulation of TM/CAM practices;
- create a stronger evidence base on the safety, efficacy and quality of the TM/CAM products and practices;
- ensure availability and affordability of TM/CAM including essential herbal medicines;
- promote therapeutically sound use of TM/CAM by providers and consumers; and
- document traditional medicines and remedies.

In many parts of the world, policy-makers, health professionals, and the public have been wrestling with questions about the safety, efficacy, quality, availability, preservation, and rational use of TM/CAM and the further development of this type of health care. It is therefore timely for WHO to define its own role in TM/CAM by developing a strategy on these issues. WHO's Traditional Medicine Strategy 2002-2005 states, "Traditional, complementary, and alternative medicine attract the full range of reactions—from being very enthusiastic to being skeptical. But use of traditional medicine (TM) remains widespread in developing countries, while use of complementary and alternative medicine (CAM) is increasing rapidly in developed countries."

Africa's traditional healers can therefore be most instrumental in preventing HIV and AIDS, for example. These traditional specialists treat most cases of sexually transmitted diseases (STDs), and experts believe that STDs are major co-factors in the spread of HIV. Secondly, in developing countries with poor infrastructure, these healers are located in nearly every rural village setting as well as in the busy urban areas. Therefore, they are a blessing to Africa's cash-strapped health ministries, since they are already "on the ground" and can contribute to solving the continent's diverse health problems.

However, despite their knowledge and popularity, traditional healers are usually given a raw deal by many Western medical experts. True collaboration between these two groups is now imperative. True collaboration, however, requires a measure of respect for indigenous medicine and African culture generally. It also requires the shedding of stereotypes of African traditional healers. Most important of all is the search for a common ground between Western biomedicine and traditional healers and building upon that common ground. Combining forces with healers to combat, for example, HIV/AIDS, and other diseases, and promote public health makes good sense.

Millions of dollars in development aid have been wasted on failed public health programs, and many lives could have been saved if only Western donors and policy-makers had a little more faith in Africa's traditional ways and gave Africans a real say in contributing to policy design and implementation. Increasingly, many more people are beginning to realize that health for all cannot be achieved without meaningful contribution of traditional healers.

In the United States and many other developing countries, use of CAM is gathering momentum as more and more people realize the value of these alternate forms of therapy. Take chronic pain, for example;

in the United States, many patients realize that they may become resistant to many medical treatments, and they can cause serious problems and side effects. Thus many patients are turning to CAM. As a result, America's National Institutes of Health's (NIH's) National Center for Complementary and Alternative Medicine (NCCAM) and its collaborators have been assessing and researching (including conducting clinical trials) the effectiveness of various CAM therapies; for example, acupuncture, massage, spinal manipulation, progressive relaxation, yoga, prolotherapy, herbal medicine, and Tai chi. They are also building an evidence base on the efficacy and safety of CAM methods for the treatment of chronic pain. Furthermore, studies on how synthetic drugs interact with herbal medicine is being conducted, and recent findings show that even grapefruit juice can interact with certain types of medication, including some calcium channel blockers, statins, benzodiazepines, and other neurological and psychiatric medications. St. John's wort, a herbal medicine, has been shown to interact with a large number of medications. In most cases, it decreases the effectiveness of the medication; in other cases, however, St. John's wort may increase a medication's effectiveness.

Africa in general and Sierra Leone in particular, can follow suit by exploiting what therapies work and eliminating those that do not work or are unsafe.

This book, although not exhaustive, will contribute to the existing knowledge on traditional medicine and medicinal plants in Sierra Leone. It includes amongst other things over 100 different medicinal uses and over 200 different plant species; it also explains how the drugs are prepared and used. It is hoped that it will be an enhancement to biomedical health care and will further the discussion on how to develop a sustainable model for incorporating traditional medicine into the national health care delivery system in Sierra Leone, as is being done in other African countries. In addition, it will promote further research in medicinal plants including their chemistry, biological activity, pharmacology and their conservation; encourage research and development of crude extracts and isolated compounds; and serve as a base for further research into new drug development.

The book is arranged in four parts:

Part 1 gives an introduction to traditional medicine.

Part 2 is a list of the plants classified according to the ailments or diseases cured.

Part 3 contains the medicinal plant species in alphabetical order, by family, with their local names (Deighton 1957; Fyle and Jones 1980); botanical names (standard scientific binomial nomenclature); the plant location; and methods of preparation of the drugs and their uses.

The common names are listed in an index on page 163

Part 4 gives a brief discussion of the chemical analysis and biological activity research of selected Sierra Leonean plants.

All attempts were made to preserve the information obtained from the ethnobotanical interviews of the herbalists in as close a form as they were conveyed to us, with little editing or interpretation. It is their information, thus it was reported as such in this book.

The letters used refer to the area from which the plants were collected (G = Gloucester Village; P = Pujehun district; H= Hastings Village; Ko = Kono district; PL = Port Loko district; Ma = Makeni district; Ke = Kenema district) or the local name of the plant in various languages (M = Mende; K = Krio; T = Themne; S = Soosoo; L = Loko; Ko = Kono).

Dr. MacFoy does not recommend self-diagnosis or self-medication. Please see the disclaimer for more information.

Disclaimer: The herbal information in this book is intended for educational purposes only. It is not the intention of the author to advise on health care. Please see a medical professional about any health concerns you have, as the book is not intended to prevent, diagnose, treat, or cure any disease.

This information is intended as an introduction to how medicinal herb plants are used. I am not a herbalist, and I cannot prescribe what herbs are right for you. If you use herbs, do so responsibly. Consult your doctor about your health conditions and use of herbal supplements. Herbs may be harmful if taken for the wrong conditions, used in excessive amounts, combined with prescription drugs or alcohol, or used by persons who don't know what they are doing. Just because a herbal remedy is natural does not mean it is safe. There are herbs that are poisonous.

PART I.
Introduction to Traditional Medicine

What Is Health?

The World Health Organization (WHO) defines health as "a state of complete physical, mental, and social well-being and not merely the absence of disease or infirmity."

Health is therefore essential for all citizens of the world, to permit them to live socially and economically productive lives, and peoples of the world have been striving to achieve this state in one way or another over the years.

What Is Traditional Medicine?

WHO defines traditional medicine (TM) as the sum total of the knowledge, skills, and practices based on the theories, beliefs, and experiences indigenous to different cultures, whether explicable or not, used in the maintenance of health as well as in the prevention, diagnosis, improvement, or treatment of physical and mental illness.

The specialists include herbalists, traditional psychiatrists, traditional pediatricians, traditional birth attendants (TBA), bone setters, occult practitioners, herb sellers, and general practitioners. They are more readily available, accessible, and approachable than orthodox allopathic physicians, while their services are much more affordable than modern medical facilities, and they play a very important role in the health of many communities.

According to WHO, traditional medicine is a comprehensive term used to refer both to TM systems (such as traditional Chinese medicine, Indian ayurveda, and Arabic unani medicine) and to various forms of indigenous medicine. TM therapies include medication therapies, if they involve use of herbal medicines, animal parts or minerals, and nonmedication therapies, if they are carried out primarily without the use of medication, as in the case of acupuncture, manual therapies, and spiritual therapies. In countries where the dominant health care system is based on allopathic medicine, or where TM has not been incorporated into the national health care system, TM is often termed "complementary," "alternative," "ethnomedicine," or "non-conventional medicine" as opposed to biomedicine or allopathic medicine (WHO, 2013)

Brief History of Traditional Medicine

No one knows for sure when humans began using herbs for medicinal purposes. My guess is that they must have done so from the evolution of man in Africa; given that other primates were already using plants for that purpose (http://www.herbalremediesinfo.com/history-of-herbal-medicine.html), and many other animals including carnivores like dogs, feed on certain plants when they are ill.

Historical evidence, however, shows that the Middle East, including the cradle of ancient civilization, Egypt, and Iraq could also be the origin of traditional medicine (Finch 1990). Knowledge of herbal medicine is well documented from Mesopotamia (Iraq) (2900 BC), Egypt (1500 BC), China (1100 BC), India (1000 BC), Greece (300 BC), Rome (100 BC), and the Bible and other religious documents.

Over the years, traditional medicine has evolved in all countries in all regions of the world, developed and developing countries included (Soforowa 1982).

In developing countries, use of traditional medicine remains widespread. According to WHO, up to 80 percent of the population in Africa use TM, while use of complementary and alternative medicine (CAM) is increasing rapidly in developed countries (in the United States, for instance, 42 percent of the population has used CAM at least once; in Australia, 48 percent; Canada, 70 percent; France, 75 percent; and Belgium, 30 percent). The annual expenditure of CAM in the United Kingdom is estimated at $2,300 million. This popularity in use, according to WHO, is due to people worrying about side effects of chemical drugs; and the belief that CAM offer a gentler approach to taking care of their health problems.(WHO, 2002)

Traditional African medicine is a holistic discipline involving the use of indigenous herbs combined with aspects of African spirituality.

People rely on traditional medicine for their basic health needs. In some cases, traditional medicine is the only health care service available. It is accessible and affordable to many people on the continent. Therefore, this significant contribution of traditional medicine as a major provider of health care services in Africa cannot be underestimated. In Uganda, for example, the ratio of traditional healers to the population is between 1:200 and 1:400, while the ratio for allopathic doctors is 1:20,000, and most are located in the urban areas.

Studies on traditional medicine and medicinal plants are being conducted in almost every part of the world, from Europe, China, India, South America, and the United States, to Africa, Australia, and New Zealand. In some countries, TM is even incorporated into the health delivery system (Duke 1998; Soforowa 1982; WHO, 2013)

The earliest recorded form of herbal medicine was possibly the marshmallow root, which is a common grass chewed for settling an upset stomach; presumably, that plant is eaten for that reason by our closest evolutionary relatives, chimpanzees and bonobos (Herbal Medicine, 2013).

Many other plant species have been found to be eaten by wild chimpanzees to treat and protect themselves from injury and illnesses, such as *Aspilia* and *Ficus* species (Newton 1991). Research on these and other genera further indicates that Neanderthals used medicinal plants to combat illnesses (Lietava 1992).

Since the dawn of civilization, humans utilized plants for their medicinal and edible value. Using trial and error, they distinguished between beneficial and poisonous plants. People also observed that in large quantities, even medicinal and edible plants can be poisonous. They also learned about the usefulness of plants by observing sick animals, which often use certain plants that they usually ignore when they are in good health.

Figure 1—*Aspilia africana*

Today, this approach, called zoopharmacology, is one of the methods used by scientists to isolate active compounds from medicinal plants used by animals, with a view to developing new drugs.

Throughout the world, there is a vast reservoir of plants with multifarious medicinal and other properties. This untapped source consists of plants that are antidiabetic, abortifacients, insect repellants, attractants, and pesticides, to name a few. Over the years, people have used medicinal plants for the treatment of various diseases. These traditional practitioners (herbalists) have developed a store of knowledge concerning the therapeutic value of these herbs. In fact, their entire pharmacopoeia is in their heads, thus quite a lot of useful information is lost when a traditional healer dies, even though attempts are usually made to pass on this information to their children.

In view of the above, however, there is now a worldwide upsurge in the interest in traditional medicine in many countries, and organizations such as WHO, UNIDO, UNESCO, UNU, Commonwealth Secretariat, and the African Union have already recognized the use of these plants and are thus supporting, promoting, or developing traditional forms of medicine. Some institutions in the United States now have departments addressing and researching and practicing integrative medicine, such as the NCCAM, Duke University's integrative medicine department, and integrative centers at Harvard, Johns Hopkins, and Columbia University (Harvard's facility is the Harvard Medical School Division for Research and Education in Complementary and Integrative Medical Therapies). In terms of usage, herbal preparations are widely used in the United States, Germany, France, and the UK, among others, running into billions of dollars each year.

Also, most governments in other developed and developing countries, including African countries such as Ghana, Nigeria, Mali, and Zimbabwe, have produced publications on their medicinal plants usage in the form of pharmacopeias; these governments are researching these plants, and many (including Sierra Leone) have gone further by recognizing and training traditional birth attendants (Williams 1979; West 1981).

Some countries have institutionalized traditional medicine into their health care systems (e.g. Mali and Senegal have integrated TM into their health care delivery system in a more advanced way by collaboration between traditional and conventional practitioners (WHO-AFRO, 2010). Others, like Nigeria, are producing drugs from wild or cultivated plants (Natural Standard, 2013)

This renewed interest in medicinal plants has been necessitated by the rising costs of drugs and the already proven efficacy of crude plant extracts and purified compounds in many countries, including China,

India, and Mexico (Dixon 1981; Kokwaro 1976). After all, it must be remembered that many conventional drugs produced in developed countries today were originally derived from raw plant material (more than 30 percent of modern medicines are derived directly or indirectly from medicinal plants). For example, aspirin is derived from the willow bark (Vickers and Zollman 1999); vincristine and vinblastine (antitumor drugs) come from Madagascar/rose periwinkle; digitalin (a heart regulator) comes from *Digitalis* and ephedrine (a bronchodilator used to decrease respiratory congestion) is derived from *Ephedra*; while atropine comes from *Atropa belladonna*; and cocaine, codeine, quinine, papain, and over one hundred other drugs were originally discovered through research on plants (See Table1), including the antimalaria drug artemisinin from *Artemisia annua* L., a plant used in China for almost 2000 years (WHO, 2013).

The Kenyan plant *Maytenus buchananii* produced an active compound in the United States, called Maytansine, which can suppress the growth of cancer-causing organisms, while *Maytenus obscura* and *M. buchananii* had been used by traditional healers in Nairobi for the treatment of cancer patients already discharged from hospitals as incurables (Kokwaro 1976).

However, in subsequent early clinical studies, maytansine was found to produce severe toxicity and unimpressive antitumor effects, which lead to the termination of its further development. Its side effects and lack of tumor specificity have prevented successful clinical use. But recently, antibody-conjugated maytansine derivatives have been developed to overcome these drawbacks (Lopus et al. 2010).

Catharanthus roseus (Madagascar/rose periwinkle) originally used for diabetes is now also used for blood cancer, ostensibly because of the two alkaloids, vinblastine and vincristine. Gossypol, which comes from cotton, is used for bronchitis in traditional medicine in China and now shows promise as a male contraceptive, and Nigerian *Zanthoxylon zanthoxyloides,* which contains p-amino-benzoic acid, other benzoic acids, and zanthoxylol, has antisickling activity (Soforowa 1982).

Taxol, which is also used in cancer chemotherapy, originates from *Taxus brevifolia,* the American Pacific yew tree. It is used in the treatment of many forms of cancer (e.g., breast, ovarian, prostate, lung, bladder, esophageal, and melanoma, as well as other types of solid tumor cancers). It has also been used in Kaposi's sarcoma, a cancer that causes lesions (abnormal tissue) to grow under the skin; in the lining of the mouth, nose, and throat; or in other organs, especially in AIDS patients (NCI, (2013).

More recently in Nigeria, a drug based on the ethanol/water extract of Fagara (*Zanthoxylon zanthoxyloides*) leaves, *Piper guineense* seeds, *Pterocarpus osum* stem, *Eugenia caryophyllus* fruit, and *Sorghum bicolor* leaves has been commercialized in the United States under the name Nicosan. This drug has been credited with alleviating sickle cell disease symptoms and reducing the number of sickle cell crises (Natural Standard, 2013). Unfortunately, this drug is no longer available in the United States at the present time.

In addition, jobelyn, Nigerian extracts from *Sorghum bicolor,* have now been commercialized as antioxidant and anti-HIV/AIDS medications (http://www.afritradomedic.com/), and the African cherry (*Prunus africanus*), a prostate treatment for benign prostatic hyperplasia, is used to increase libido, for treating cancer, and as an anti-inflammatory agent. The bark or processed extracts of this plant from countries such as Cameroon, Zaire, Kenya, and Madagascar are exported to Europe for preparation of the drugs (Stewart 2003).

Over the past few years, *Moringa oleifera* (the tree of life) (figure 2), a plant originally from India, has become very popular in many countries including Sierra Leone, where it is already being cultivated. It is thought to have nutritional and medicinal values that have the propensity to cure and prevent over three hundred diseases, including hypertension and diabetes. Advocates say that it can prevent diseases and malnutrition and can even boost development by creating job opportunities. It is thought to contain seven times the Vitamin C found in oranges, four times the calcium in milk, four times the vitamin A in carrots, and three times the potassium in bananas. According to an article by Adevu (2011) of the United Method Church on Relief (UMCOR) the plant contains some forty-six antioxidants and is very rich in phytonutrients, which flush toxins from the body, purify the liver, and bolster the immune system. It is regarded as a cure-all, and many people in some communities are no longer visiting the local clinics, as they are now either using *Moringa* oil or teabags, or sprinkling the powder on their daily meals.

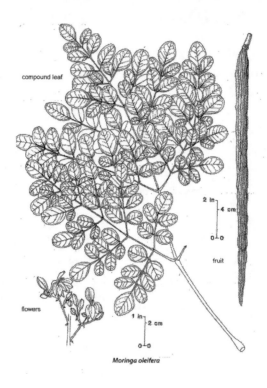

compound leaf

2 in — 4 cm

0 — 0

fruit

flowers

1 in — 2 cm

0 — 0

Moringa oleifera

Figure 2—*Moringa oleifera*

The tables below document the widespread use and medicinal potential of various plant species.

Table 1 is a list of drugs derived from plants (Fabricant and Farnsworth, 2001) while Table 2 shows some potential drug plants from African forests, and Table 3 lists some African medicinal plants in the world market (Soforowa, 1993b).

Table 1 is not a full, comprehensive list of all patented drug names and herbal drugs sold in all countries, but it is quite representative. The chemical or drug names are given together with their action or clinical use and the plant names from which the drugs were developed (Fabricant and Farnsworth 2001)

Table 1. List of Drugs and Chemicals Derived from Plants

Drug/Chemical	Action/Clinical Use	Plant Source
Acetyldigoxin	Cardiotonic	Digitalis lanata
Adoniside	Cardiotonic	Adonis vernalis
Aescin	Anti-inflammatory	Aesculus hippocastanum
Aesculetin	Antidysentery	Frazinus rhychophylla
Agrimophol	Anthelmintic	Agrimonia supatoria
Ajmalicine	Circulatory disorders	Rauvolfia serpentina
Allantoin	Vulnerary	Several plants
Allyl isothiocyanate	Rubefacient	Brassica nigra
Anabesine	Skeletal muscle relaxant	Anabasis sphylla
Andrographolide	Baccillary dysentery	Andrographis paniculata
Anisodamine	Anticholinergic	Anisodus tanguticus
Anisodine	Anticholinergic	Anisodus tanguticus
Arecoline	Anthelmintic	Areca catechu
Asiaticoside	Vulnerary	Centella asiatica
Atropine	Anticholinergic	Atropa belladonna
Benzyl benzoate	Scabicide	Several plants
Berberine	Bacillary dysentery	Berberis vulgaris
Bergenin	Antitussive	Ardisia japonica
Betulinic acid	Anticancerous	Betula alba
Borneol	Antipyretic, analgesic, anti-inflammatory	Several plants
Bromelain	Anti-inflammatory, proteolytic	Ananas comosus
Caffeine	CNS stimulant	Camellia sinensis
Camphor	Rubefacient	Cinnamomum camphora

Camptothecin	Anticancerous	Camptotheca acuminata
Catechin	Haemostatic	Potentilla fragarioides
Chymopapain	Proteolytic, mucolytic	Carica papaya
Cissampeline	Skeletal muscle relaxant	Cissampelos pareira
Cocaine	Local anaesthetic	Erythroxylum coca
Codeine	Analgesic, antitussive	Papaver somniferum
Colchicaine amide	Antitumor agent	Colchicum autumnale
Colchicine	Antitumor agent, antigout	Colchicum autumnale
Convallatoxin	Cardiotonic	Convallaria majalis
Curcumin	Choleretic	Curcuma longa
Cynarin	Choleretic	Cynara scolymus
Danthron	Laxative	Cassia species
Demecolcine	Antitumor agent	Colchicum autumnale
Deserpidine	Antihypertensive, tranquilizer	Rauvolfia canescens
Deslanoside	Cardiotonic	Digitalis lanata
L-Dopa	Anti-Parkinsonism	Mucuna sp
Digitalin	Cardiotonic	Digitalis purpurea
Digitoxin	Cardiotonic	Digitalis purpurea
Digoxin	Cardiotonic	Digitalis purpurea
Emetine	Amoebicide, emetic	Cephaelis ipecacuanha
Ephedrine	Sympathomimetic, antihistamine	Ephedra sinica
Etoposide	Antitumor agent	Podophyllum peltatum
Galanthamine	Cholinesterase inhibitor	Lycoris squamigera
Gitalin	Cardiotonic	Digitalis purpurea
Glaucarubin	Amoebicide	Simarouba glauca
Glaucine	Antitussive	Glaucium flavum
Glasiovine	Antidepressant	Octea glaziovii
Glycyrrhizin	Sweetener, Addison's disease	Glycyrrhiza glabra
Gossypol	Male contraceptive	Gossypium species
Hemsleyadin	Bacillary dysentery	Hemsleya amabilis
Hesperidin	Capillary fragility	Citrus species
Hydrastine	Hemostatic, astringent	Hydrastis canadensis

Hyoscyamine	Anticholinergic	Hyoscyamus niger
Irinotecan	Anticancer, antitumor agent	Camptotheca acuminata
Kaibic acid	Ascaricide	Digenea simplex
Kawain	Tranquilizer	Piper methysticum
Kheltin	Bronchodilator	Ammi visaga
Lanatosides A, B, C	Cardiotonic	Digitalis lanata
Lapachol	Anticancer, antitumor	Tabebuia sp.
a-Lobeline	Smoking deterrent, respiratory stimulant	Lobelia inflata
Menthol	Rubefacient	Mentha species
Methyl salicylate	Rubefacient	Gaultheria procumbens
Monocrotaline	Antitumor agent (topical)	Crotalaria sessiliflora
Morphine	Analgesic	Papaver somniferum
Neoandrographolide	Dysentery	Andrographis paniculata
Nicotine	Insecticide	Nicotiana tabacum
Nordihydroguaiaretic acid	Antioxidant	Larrea divaricata
Noscapine	Antitussive	Papaver somniferum
Ouabain	Cardiotonic	Strophanthus gratus
Pachycarpine	Oxytocic	Sophora pschycarpa
Palmatine	Antipyretic, detoxicant	Coptis japonica
Papain	Proteolytic, mucolytic	Carica papaya
Papavarine	Smooth muscle relaxant	Papaver somniferum
Phyllodulcin	Sweetener	Hydrangea macrophylla
Physostigmine	Cholinesterase inhibitor	Physostigma venenosum
Picrotoxin	Analeptic	Anamirta cocculus
Pilocarpine	Parasympathomimetic	Pilocarpus jaborandi
Pinitol	Expectorant	Several plants
Podophyllotoxin	Antitumor anticancer agent	Podophyllum peltatum
Protoveratrines A, B	Antihypertensives	Veratrum album
Pseudoephedrine	Sympathomimetic	Ephedra sinica

Pseudoephedrine, nor—	Sympathomimetic	Ephedra sinica
Quinidine	Antiarrhythmic	Cinchona ledgeriana
Quinine	Antimalarial, antipyretic	Cinchona ledgeriana
Quisqualic acid	Anthelmintic	Quisqualis indica
Rescinnamine	Antihypertensive, tranquilizer	Rauvolfia serpentina
Reserpine	Antihypertensive, tranquilizer	Rauvolfia serpentina
Rhomitoxin	Antihypertensive, tranquilizer	Rhododendron molle
Rorifone	Antitussive	Rorippa indica
Rotenone	Piscicide, Insecticide	Lonchocarpus nicou
Rotundine	Analgesic, sedative, tranquilizer	Stephania sinica
Rutin	Capillary fragility	Citrus spp.
Salicin	Analgesic	Salix alba
Sanguinarine	Dental plaque inhibitor	Sanguinaria canadensis
Santonin	Ascaricide	Artemisia maritma
Scillarin A	Cardiotonic	Urginea maritima
Scopolamine	Sedative	Datura species
Sennosides A, B	Laxative	Cassia species
Silymarin	Antihepatotoxic	Silybum marianum
Sparteine	Oxytocic	Cytisus scoparius
Stevioside	Sweetener	Stevia rebaudiana
Strychnine	CNS stimulant	Strychnos nux-vomica
Taxol	Antitumor agent	Taxus brevifolia
Teniposide	Antitumor agent	Podophyllum peltatum
Tetrahydrocannabinol (THC)	Antiemetic, decrease ocular tension	Cannabis sativa
Tetrahydropalmatine	Analgesic, sedative, tranquilizer	Corydalis ambigua
Tetrandrine	Antihypertensive	Stephania tetrandra
Theobromine	Diuretic, vasodilator	Theobroma cacao
Theophylline	Diuretic, bronchodilator	Theobroma cacao and others
Thymol	Antifungal (topical)	Thymus vulgaris
Topotecan	Antitumor, anticancer agent	Camptotheca acuminata

Trichosanthin	Abortifacient	Trichosanthes kirilowii
Tubocurarine	Skeletal muscle relaxant	Chondodendron tomentosum
Valapotriates	Sedative	Valeriana officinalis
Vasicine (peganine)	Cerebral stimulant	Vinca minor
Vinblastine	Antitumor, Antileukemic agent	Catharanthus roseus
Vincristine	Antitumor, Antileukemic agent	Catharanthus roseus
Yohimbine	Aphrodisiac	Pausinystalia yohimbe
Yuanhuacine	Abortifacient	Daphne genkwa
Yuanhuadine	Abortifacient	Daphne genkwa

Table 2. Some Potential Drug Plant Candidates from African Forests

Plant species	Part used	Indications	Constituents
Phytolacca dodecandra	Leaf	Schistosomiasis	Triterpenoid saponins
Tetrapleura tetraptera	Fruit	Schistosomiasis	Aridanin
Azadiracta indica	Leaf/bark	Fever/Malaria	Gedunin
Cryptolepis sanguinolenta	Root	Hypertension	Cryptolepine
Zanthoxylum zanthoxyloides	Root	Sickle cell anaemia	Phenolic acids
Bridelia ferruginea	Bark	Diabetes	-
Cajanus cajan	Seed	Sickle cell anaemia	Phenylalanine
Senna podocarpa	Leaf/pod	Constipation	Anthraquinones
Senna alata	Leaf/Pod	Constipation/Eczema	Anthraquinones
Datura metel	Aerial part	Ulcer and dysmenorrhea	Hyocyamine
Momordica charantia	Leaf/fruit	Diabetes	Charantin
Euphorbia hirta	Aerial part	Asthma and dysentery	-
Aloe schweinfurthii	Leaf	Skin infections, burns	Anthraquinones/Enzymes

Table 3. Some African Medicinal Plants in the World Markets

Plant Species	Major Indications	Source(s)
Ancistrocladus abbreviatus	Anti-HIV	Cameroon & Ghana
Corynanthe pachycerus	Male stimulant	Ghana
Physostigma venenosum	Ophthalmia	Nigeria, Ghana
Prunus africana	Prostrate gland hypertrophy	Cameroon, Kenya, Madagascar
Rauwolfia vomitoria	Psychiatry/Hypertension	Nigeria, Zaire, Rwanda, Mozambique
Strophanthus gratus.	Cardiotonic	West African Countries
Voacanga africana	Antimicrobial	Cote d'Ivoire, Ghana, Cameroon, Nigeria
Albizzia adianthifolia	Gum	South Africa

Each plant species can be regarded as a distinct and complicated biochemical factory, producing many compounds with vast potentialities for medicinal use. However, it must always be borne in mind that the plant itself varies in its potential to biosynthesize its multifarious chemical constituents. For instance, there are seasonal and other variations due to the age, source, stage of maturity, part of the plant used, time of the day harvested and ecology in which the plant is grown. Some plants contain higher concentrations of the active ingredients in the morning than at night and vice versa. Others contain more ingredients when the yellow dead leaves on the ground are used instead of the healthy green leaves on the tree (Soforowa, 1982)

There is already a lot of evidence to support the view that to a great extent, most plants used in traditional medicine contain active chemical constituents responsible for curing the ailments of many patients. These chemicals include alkaloids, tannins, flavonoids, saponins, and essential oils. They are secondary metabolites biosynthesized by plants as defense mechanisms to fight off pathogens, pests, and herbivores (MacFoy 1988; 1992). Some of these plants can be powerful poisons, such as *Mareya micrantha*, hence the need for expertise and great caution in its use. However, the possibility that some plants only conform to the doctrine of signatures (that is, "an observable feature of a plant indicates its utility") should not be neglected. For example, by the doctrine of signatures, a plant which resembles the liver is used to cure liver problems; the yellow extract from *Harungana madagadcariensis* plant is used to cure yellow fever.

Other plants may just exhibit the placebo effect or placebo response. This is a phenomenon in which an inactive substance can sometimes improve a patient's condition simply because the person has the expectation that it will be helpful. This expectation plays a potent role in the placebo effect. The more a person believes they are going to benefit from a treatment, the more likely it is that they will experience a benefit.

There are various forms of traditional medicine in Africa:

1. Symbolism: This is when a particular quality is transferred from one site to another, usually by touch or even miles away. In the latter case, the magician or jujuman simply mentions the name of the person, stating the person is struck by an illness or death. An example of contagious or sympathetical magic involves the prescription of the heart of a lion for weakness or a bone treated with a magical portion and left on the road to prick the person for which it has been put.
2. Cupping: The medicine man sucks the affected part of the body and then spits out a piece of foreign object (e.g., worms, small horns, etc.) as evidence that he has drawn out the disease.
3. Removal of evil: The medicine man removes the evil spirit causing the illness from the patient and transfers it to an animal which is then driven into the forest, or the illness is left on the road to be picked up by a passerby.
4. Scarification: The medicine man makes pairs of linear incisions about a half-inch long on the site of the pain or disease, and powdered herbs are rubbed into the incision and tied. This method can also be used to provide protection.
5. Wearing of charms: This takes the form of an amulet or a piece of wood or other material suspended by a string around the neck (usually adults) or waist (usually children). Charms are usually

intended for protection against evil spirits, ill luck, or bad health, and for strength (Soforowa 1982).

Spirituality in African Medicine

As in most religions, African traditional religion divides the cosmos into two realms—the spiritual world and the natural world. In the spiritual world reside the supernatural forces (e.g. the various gods, good as well as evil), ancestors, and witches. The natural world is made up of humans, animals, and plants. These two worlds, it is believed, are interconnected with and interdependent on each other. Many African countries use the age-old, holistic, non-Western approach to medicine in addition to orthodox, allopathic western medicine. This latter treats diseases with chemicals that produce effects which are contrary (in opposition) to those of the disease being treated.

Allopathic medicine therefore treats symptoms. On the other hand, the old-age holistic method is more concerned with identifying the causes of illness and disease, as well as in removing symptoms so as to restore balance in the biological system. In addition, there are spiritual dimensions as well, that are claimed to restore holistic balance in the patient. The healer will want to find out the spiritual causes of the diseases (if there are any) as well as restoring spiritual balance in the patient (if necessary). The healer uses incantations and divination to diagnose the causes of the illness.

Thus traditional healers in Africa use both scientific and nonscientific subjective knowledge. These two major paradigms, the biomedical and spirit-inspired traditions, are well documented in the literature (Soforowa 1982). The former uses plants, many of which have been shown by modern scientific methods to be effective. This empirical knowledge has been developed over many years through trial and error, experimentation, and systematic observation. The metaphysical, subjective, nonscientific knowledge includes various spirits which are believed to be a factor in health by purportedly playing a role between God and the living. In this system, disease can be caused by either natural or supernatural causes, and divination is used to ascertain which is which. Illnesses caused by natural causes will require herbal and pharmacologic remedies, whereas those believed to be caused by supernatural forces will require the offering of sacrifice, use of talismans and amulets, or recitation of incantations. Many cultures in Africa, for example the Yoruba society, rely on both for restoring physical as well as spiritual balance (Soforowa 1982).

Sierra Leone: The Country

Sierra Leone is a small country with a total area of 71,740 square kilometers (27,699 square miles), on the west coast of Africa. Its immediate neighbors are Guinea on the north and northeast and Liberia on the southeast and on the south and southwest, it is bounded by the Atlantic Ocean. Though relatively small, Sierra Leone is endowed with natural riches that include mineral resources and biodiversity, although unfortunately the rate of biodiversity loss has become phenomenal in recent times. The capital and largest city in Sierra Leone is Freetown (Fyle 2011; Allie 1990).

Figure 3—Map of Sierra Leone

Figure 4—Map of Sierra Leone

The area now called Sierra Leone was probably first occupied by the Limbas, and by the time the Portuguese traders first came in 1462, there were communities on the coast such as the Baga, Bullum, Krim, and Vai, all together called the Sapes. In the northwest were the Themne and Loko, and in the north, the Limba. The Banta were in the southwest, while the Kissi and Kono were in the east. Invasions of the country by the Mane/Mani (a group of Mandinka from the Mande empire) changed the communities to some extent by influencing the cultural distribution of Sierra Leone. The Mendes are believed to be descendants from the Mane, originally from the Liberian interland, and the Themne claim to have come from Futa Jallon, which is present day Guinea. The Krios are an integration of freed enslaved Africans—the original settlers from Britain in 1787 (the black poor); Nova Scotians from America; Maroons from Jamaica (escaped enslaved Africans); Recaptives (liberated Africans), and people from the various local ethnic groups. Most of the peoples of Sierra Leone are Christians or Muslims, or belong to an indigenous African religion, mostly animist (Fyle 2011; Allie 1990).

Sierra Leone was a colony of Britain until 1961, when it gained its independence, and in 1971, it became a republic. It has a tropical climate with a dry season and a rainy season. The vegetation is diverse and includes pockets of tropical rain forests as forest reserves, savannah grasslands, and swamplands (Cole 1968).

According to a CIA report, the estimated population of Sierra Leone in July 2013 was 5,612,685; the estimated life expectancy at birth was 59.56 years for women and 54.47 years for men. The infant mortality rate is also high at 74.95 infant deaths per 1,000 live births. Medical personnel and facilities are grossly inadequate (CIA Report 2013). The Government's Ministry of Health and Sanitation (MOH) is responsible for providing quality affordable and accessible health care services to the people of Sierra Leone. This is complemented by the private sector and nongovernmental organizations. Nevertheless, there is still a lack of adequate health care facilities in the country, especially after the eleven-year civil war, which ended in 2002. Traditional medicine now forms part of the primary health care system in the country.

In 2010, the government started implementing a free healthcare policy for pregnant women, lactating mothers, and children under five. This policy abolished fees paid for medical attention and provided drugs and treatments free of charge in every public health facility in the country. Around 1.2 million mothers and children were expected to benefit from the new policy in 2010-2011 alone. The program is being supported by UNICEF, the UK's Department for International Development, WHO, and the European Union. Nonetheless, recent findings indicate that the program is already facing challenges.

Research Prior to 1979

Before 1979, very little research on the ethnobotany and ethnopharmacology of Sierra Leone had been carried out, although medicinal plants were widely used, especially in the rural areas. However, it was common knowledge that several plant species cured diseases and illnesses; for example, crude aqueous extracts (called *agbo* in Krio) of stinking weed (*Cassia occidentalis*) is commonly believed to lower blood sugar levels in diabetics, pawpaw (*Carica papaya*) is used to heal wounds, apparently due to the presence of the protein papain, and *Cassia siberiana* (called *Gbangba* in Mende) is good for malaria.

Some phytochemical studies had however been reported by various workers; for example, Taylor-Smith and Hanson at the chemistry department of Fourah Bay College, University of Sierra Leone (Hanson 1977; Taylor-Smith 1966). In late 1979, when the author joined the university as the biochemist/biologist, he initiated research with his students in the biology department on ethnobotanical studies and plant collection of medicinal plants, natural dyes, and other useful plants of Sierra Leone, in addition to preliminary biochemical and biological activity studies. This was partly with a view to producing a pharmacopeia and a compilation of other useful plants of the Sierra Leone flora. It was thought absolutely necessary in realization of the rapid loss of plant species as a result of massive deforestation and other disasters. The impetus was to document this information which was also being lost since they still remained unrecorded, as the herbalists and older residents were dying without passing on this indigenous knowledge orally to their children as they used to do, and to also scientifically authenticate the claims of the herbalists.

Ideally the botanists, taxonomists, and pharmacognosists should first correctly identify and classify the plants, the importance of which cannot be overemphasized; the chemists/phytochemists, biochemist pharmacologists, and microbiologists should then work together in the quest for knowledge on the medicinal value of these plants. So far this has not been feasible in any systematic and comprehensive way for lack of facilities and qualified personnel.

Research after 1979

Although some ethnobotanical research work has been conducted in many West African countries (Ayensu 1978; Dalziel 1937; Oliver-Beaver, 1986; Soforowa 1982), the work reported here is the most recent documented and comprehensive (but by no means exhaustive) field study on medical ethnobotany in Sierra Leone, which we hope will contribute to the existing knowledge in this field. The research was carried out starting in 1980 in villages of Pujehun district in the Southern province, in Gloucester Village in the mountain district in the Western area of Sierra Leone, in Hastings Village on the outskirts of Freetown, in Kono district in the eastern province, in Kenema in the eastern province, and Port Loko district in the Northern Province. They form part of a study of the whole of Sierra Leone. Preliminary results obtained in Gloucester Village and Pujehun district have already been published in my 1983 book, *Medicinal Plants of Sierra Leone*, and this second edition (which is a lot more comprehensive) incorporates this first edition and results obtained from the other regions of the country in addition to chemistry and biological activity studies. Since the first edition was published, a number of reports cataloguing some medicinal plants have been published by other researchers, such as Barnish (1992) and Turay (1997).

Long before the advent of Europeans, Sierra Leoneans had been using plants to address their health problems arising from the various diseases in the country, and they continue to do so up to this date. However, on arrival, the Europeans usually discouraged the use of TM in their colonies, even though it still continued. Regardless, even during the slave trade, this knowledge was carried over to the West Indies and to the Americas (e.g. Brazil and South Carolina) by the enslaved Africans, who continued to practice traditional medicine (MacFoy, unpublished; Mitchel 1999; Pollitzer 2005).

In our studies, information was obtained by interviewing professional herbalists and older local residents using a questionnaire. The interviews were conducted throughout the country, including the western area and villages, eastern, northern, and southern provinces. Usually a chief, headman, or a local dignitary in a village is used to help us identify well-known herbalists and to facilitate their cooperation during the interviews.

In most cases, information was obtained on the local population (viz. ethnographic, ethnological, linguistic, geographical, and geological data. The vegetation was undergoing changes through the process of deforestation, and secondary forests have replaced virgin forests; nonetheless, there are still patches of dense forests in the interior.

Secret societies exist throughout the country (e.g., in Hastings Village, about 2.5 miles from the capital city, Freetown, the hunting and *"agugu* societies provide practical answers to local health problems). Each of these utilizes two types of medicines: vindictive and curative. The vindictive ones are used to increase their membership and as a punishment, while curative ones include medicinal plants species which have been found to be useful, thereby giving them some degree of power in their respective communities.

In the field, herbalists throughout the country were accompanied by a researcher to the forest, where they would point out the plants used for treating various ailments. An on-the-spot interview was then carried out using the questionnaire containing several pertinent questions relating to the ecology, botany, pharmacognosy, and other useful information on how the plants are collected and prepared; whether there are any conditions for harvesting the plants and preparing the drug; and questions relating to dosage and storage. The answers were recorded and plant samples collected for identification. Special strategies to ascertain the authenticity of the information supplied were employed, but no inconsistencies were detected. For example, the information obtained from one herbalist was cross-checked with those of others, and the whole interview was repeated after a few days. In some cases, there were claims of some supernatural influence in disease aetiology and therefore its cure, with concomitant secrecy. In these cases, the healer will consult the oracle and focus on both the physical as well as the spiritual well-being in dealing with the disease. Thus the healing methods in such cases will include metaphysical cures, ritual foods, amulets, and charms. This aspect of traditional medicine is the least understood by an outside observer.

Every herbalist was given a small financial remuneration after the interview, and the plants collected in the field were identified at the Fourah Bay College, University of Sierra Leone, Herbarium using their vegetative and reproductive structures and their local names. They were specially prepared, mounted, and stored in the herbarium as voucher specimens.

African Medicine: Magic, Sorcery, World View

Traditional practitioners use a wide variety of treatments ranging from "magic" to biomedical methods such as fasting and dieting, herbal therapies, bathing, massage, and surgical procedures. From our research, it is clear that not all traditional medicine in Sierra Leone is herbal per se; some preparations contain animal or insect parts; from frogs, bees, chameleons, snakes, goats, monkeys, and so on.

Herbal medicine is defined as the use of medicinal plants to prevent or heal disease. According to WHO, a medicinal plant is defined as any plant which contains, in one or more of its organs, substances that can be used for therapeutic purposes or which are precursors for chemotherapeutical semisynthesis. Herbal remedies have formed the basis of traditional medicine for millennia and form the root of modern pharmacology (http://www.herbalremediesinfo.com/history-of-herbal-medicine.html).

Apart from using plants as medicine, humans have always relied on plants for their food, dyes, timber, and other necessities (Cole and MacFoy 1992; MacFoy 2008).

In Sierra Leone, drugs are usually prepared by one or more of the following methods: infusion, decoction, maceration, juice, powder, steam, and poultice/cataplasm.

- Infusion: Usually hot water is poured onto plant part(s) and then filtered to obtain the liquid drug.
- Decoction: A preparation made by boiling crude vegetable drugs in water, which is then strained or filtered to obtain the liquid drug.
- Maceration: Cold water is added to the plant material and left standing for some time at room temperature before straining to obtain the liquid drug.
- Cataplasm: Fresh parts are ground into a paste or poultice and applied to the affected areas.

In addition, drugs are also prepared from plants by distillations, and the distillate used for drinking or bathing. Generally, drugs are applied either internally or externally. For the latter, the drug may be applied as a poultice, liniment, lotion, or baths, while for internal use, they may be chewed, ingested, inhaled, or taken as an enema. Solvents used include water, palm oil, palm wine, a locally distilled rum or gin called *omole* in Krio, and animal (such as goat) fat.

Our research in Sierra Leone also shows that about 90 percent of the medicinal plants are collected in the wild; only a few are cultivated (e.g., *Carica papaya, Ocimum gratissimum, Psidium guava, Cajanus cajan, Hibiscus esculentus, Phaseolus lunatus,* and *Citrus aurantifolia,* which are actually cultivated for their nutritional use, not purposely for their medicinal value). Some plants are used to cure more than one ailment. Some are used alone, while others are blended with other plants to produce the drug, apparently to increase potency or to add flavor to the drug. In some cases, palm wine (a local alcoholic beverage made from the palm tree) or *omole* (a local gin) is added to the drug according to the herbalist to increase its potency. This is not surprising, since alcohol is an organic solvent that extracts more of the active ingredients from the plant than water would.

Dosage is always an issue; the herbalists are well aware of this and give advice on dosage, though not as precisely as in allopathic medicine (e.g., a handful or a cupful of a drug for an adult and half of that for a child, as the case may be). In all cases, a small financial inducement is given to the herbalist for sharing his knowledge, while in some cases, kola nuts (*Cola nitida*) are also requested, as a small token for revealing their knowledge. In Kono district, for example, the foreign herbalists from Mali or Guinea request exorbitant amounts. Secrecy was maintained since they regard their herbal knowledge as a legacy from their parents, through patrimony, hence they hardly make their full knowledge available to others, including researchers, even though they charge a certain amount.

One herbalist for broken bones in Kenema indicated during an interview that he obtained his healing knowledge from his late father in a dream, and he will only reveal that knowledge to his son in the same way. Many traditional healers in the country consult the oracle and include both the physical body and the spiritual well-being when dealing with illness. Therefore, traditional healing methods in some areas include not only herbs but divine healings like metaphysical cures, ritual foods, amulets, and charms.

Some herbalists supplement herbs with divine metaphysical dimensions, phenomena, and methods. These herbalists usually wear amulets or charms, and chew one or two special leaves before they depart to the forests for the survey with the research investigator.

Some drugs conform to the doctrine of signature (e.g. *Costus afer* stem infusion, which produces a yellow liquid, is used for yellow fever, and so is the plant *Craterispermum laurinum*, and *Angelaca oblique* leaves, which produce a red liquid, is used for anemia).

Most of the herbalists interviewed considered a plant as medicinal from personal experience or acquired the knowledge from their ancestors, or if it satisfied the expectations of the patient and the herbalist, then it was regarded as the proper drug.

If one drug fails to cure the condition, as with allopathic medicine, another herb is used or the advice of another herbalist is sought. Some herbalists perform rituals which they claim will foretell whether a particular herb will cure the condition of a particular patient. If this ritual exercise proves negative, the patient may be rejected for treatment or referred to another herbalist.

In Sierra Leone, traditional medicine, like allopathic medicine, covers a wide range of fields, including protective medicine (immunization), curative medicine, and destructive medicine (i.e. voodoo/juju). This latter is tied to sorcery, superstition, and mysticism (Traditional African Medicine (n.d.)). The protective aspects of using herbs includes protection against snake bites, scorpions, centipedes, tarantula stings, food poisoning, witchcraft, lightning and thunder, and gun shots, and as a contraceptive. The curative aspects include healing wounds, boils (hard, poisoned swelling under the skin), heart problems, infertility, and impotence; increasing mental capability; wooing an individual into marriage; and others, while the destructive medicine includes using herbs for transmitting infectious and contagious diseases, causing food or drink poisoning, and making individuals suffer misfortune or some form of physical or mental disability.

Aguwa (2002) stated that in Mende medicine, healers called medicine men (*halei*) and secret medicine societies deal with illnesses, which can have physical or spiritual causes. The spiritual causes include individual moral deficiencies and the malevolent activities of spirits and their agents. The diviner discovers the cause of an illness or misfortune. The medicine man or healing doctor then prepares medicines and administers them. Medicines are prepared from herbs and other natural substances. Protective medicine can consist of charms and inoculation with "power substances." When medicine is prepared and consecrated, it is believed to be endowed with efficacy. Since the advent of scientific medicine and Christianity in Mende and other societies, the use of traditional medicine has been on the decline, especially in towns and cities, but still high in villages and poor arears.

Poisonous Plants

Some plants are quite poisonous and have been used in ritual murders, on poison arrows for battle, or for killing wild animals and for fishing.

Kargbo (1982) discussed several poisons from biological sources; for example, the plant *Newboulder laevis* (used as a poultice for strangulated hernia) is also used by some sorcerers to demonstrate their powers of destruction (e.g., when investigating a theft in the village, they may first expose a live chicken to the fumes of this plant, specially prepared from the roots, which then causes death of the chicken; as a result of this dramatic demonstration, the thief usually makes a confession).

The bark of the plant *Erythrophylum guineense* (sasswood or ordeal poison) is used as a fish poison by spreading the powdered bark into a stream. The fish will go into a swoon and are then collected for preparation and consumption. This same plant is also used by "witch hunters" to get rid of "witches." Preliminary toxicity studies on rats with this plant, together with *Bersama abyssinica* and *Mareya micrantha*, were conducted by Taylor-Smith and Chaytor (1966).

The beans and fruit inside the pod of *Parkia biglobossa* (locust tree) is an edible spice (*kenda*), and the bark is used for spinal injuries; however, the skin of the pod is poisonous to fish and humans.

The bile from the gall bladder of the crocodile and from an egret are both poisonous to humans, as is the concentrate from the bitter cassava variety (*Manihot esculentum*) mixed with the root of *Cassia sieberiana* (Gbangba).

A very potent poison is obtained from the fumes of suitably prepared dead human flesh, containing cadavarine and putricine from the putrifying flesh (Kargbo 1982).

Our research indicates that some drugs (e.g., the laxative *Mareya micantha*) should not be given to pregnant women, since they can act as an abortifacient. It should also not be given to children and weak patients, since it is poisonous in large doses. A few other poisonous plants are used for stunning fish (*Tephrosia voogelii*) and for killing vermin (the rat poison, *Dichapetalum toxicarium*).

Some herbs may just have the placebo effect, and since some herbalists are excellent psychologists, patients may just believe they are getting better by being convinced by the herbalist. Therefore, these particular drugs only have a psychological effect on the patient. Moreover, some unscrupulous herbalists are also able to convince patients that they are very ill or are going to die on a certain day, with the hope of extorting money from them.

Traditional Birth Attendants

This is another group that provides traditional medicine in Sierra Leone. A large number of the babies in the country are delivered by TBAs, and they use herbal medicine to a great extent in their practice (fifty-three plants species identified) (Williams 1979; Williams, unpublished list, 1988); they also have a lot of customs and traditions associated with childbearing and childrearing (Williams 1979; Davies et al. 1992; Kargbo 1986).

Since 1979, training of TBAs has been conducted by the Ministry of Health (MOH), with the aim of reducing mortality and morbidity rates in young children and mothers during pregnancy, labor, and delivery. A training manual was published in 1979 (Williams 1979). Hence, these TBAs have been integrated into the health delivery system of the country, to some extent.

TBAs are trained to perform sterile, safe deliveries and recognize the warning signs of complications of childbirth which require trained assistance. Another important role is to provide comfort to the women who are naturally anxious, given the high maternal death rate and the financial and physical problems that can occur. The fees charged by traditional birth attendants were approximately $10 to $15 for a straightforward delivery.

Many traditional birth attendants are illiterate. When they have completed their initial training, they are given a wooden "statistics" box and a certificate. Each compartment in the box has a picture, and the box is used as a method of compiling statistics. The box has five compartments, and TBAs put small stones in the appropriate section of their box to record details of the births that they attend (e.g., number of live births, number of stillbirths, and if she had to refer the mother to the local clinic).

The TBA box also has a larger wooden box that contains basic supplies for the attendants' use: surgical spirit, cotton wool, safety pins, nail clippers, razor blades, torch and batteries, towels, bandages, gauze, soap, needle and thread, apron, and plastic sheeting.

In addition, the traditional female societies (Davies et. al. 1992) and male societies also use herbal medicine in their activities (Hinzen 1986).

Faith Healing

This is defined as the recourse to divine power to cure mental or physical illness; "they shall lay hands on the sick and they shall recover" (Mark 16:18 RSV). Although greatly advocated by religious groups, faith healing is gradually appearing in medical literature, so belief in miracle cures and healing may not be wholly independent of medical science.

Faith healing holds a firm root in the use of intercessory prayers for healing. The healers believe that this does not negate the fact that God is the ultimate physician, who in turn has empowered other physicians to heal and cure diseases and infirmities. The power to cure and the gift of healing through medical science is given by God and should not be shunned. A Jewish saying is "the Lord created medicines out of the earth and the sensible man will not despise them (Hamdan 2009)

"Faith healers believe that Christians should not only believe in God, they must believe that God makes the final decision on a cure being a success or a failure, either spiritually or medically. For example, some mental illnesses are still believed by some people to be caused by evil spirits and thus cured by exorcism.

Scientists are now trying to validate the claim that prayers are effective in curing illnesses by using the scientific method, as more and more patients are requesting that their doctors and nurses use prayers as part of their treatment. Many non-Christian faiths also believe in the power of prayers to cure illnesses, and the NIH is looking into how effective prayer is in medical treatment (NCCAM Pub No: D456 2011)

Islamic Healers

These healers use Arabic Islamic verses from the Quran as a remedy or part of a remedy, together with local herbs, for treating diseases such as paralysis, mental illness, and impotence. Usually certain verses of the Quran are written on a piece of paper or on a tablet, and the ink is washed out and used with the herb. Thus this method of healing is both spiritual and herbal, involving incantations; it assumes that there are some supernatural powers being invoked to bring about the treatment (Turay 1997).

Some of these healers claim to have expertise in divination. Herbalists seem to use different techniques for divination than Mori-men (Muslim healers) and Karamokos (Muslim teachers). For instance, one healer in Makeni uses *diegge* (cowries), and in Kenema, a herbalist uses his *yorgoi* (pebbles) and *totewei* (stick), while the Mori-men and Karamokos have a preference for Muslim methods, including the *Tasabia* (rosary), *Ramla* (numbers), and *Atmera* (writing of verses from the Quran (Utas 2009)

Also in this article, one herbalist in Freetown who cures victims of sexual abuse stated that she uses herbs in combination with prayers from priests and imans to cure her patients.

In general, when seeking a cure, some patients utilize a variety of treatment options either simultaneously or one after the other, if they perceive that the first did not solve the problem.

Indigenous Knowledge and Biodiversity Conservation of Medicinal Plants

Because of the rapid rate of deforestation that is currently occurring in Sierra Leone, it is imperative that all efforts be made to prevent this massive deterioration. The Sierra Leone forests are under extreme pressure from human activities such as hunting, wood cutting, logging, mining, and the creation of new settlements.

Biodiversity conservation of useful Sierra Leonean plants has already been discussed for medicinal plants (MacFoy 1992) and for natural dye plants (MacFoy 2004; 2008).

This can be achieved by bioprospecting and involving communities in projects that will boost their livelihood, aimed at helping to alleviate poverty and ensure food security and improved health. Thus all efforts must be made to reverse this deforestation trend by the use of sustainable income-generating initiatives such as eco-tourism.

The goal should be the creation of alternative livelihoods that will promote income generation in communities, which in turn will reduce rural poverty. At a national level, the projects should include the promotion of value-added processing of Medicinal, Oil-bearing, Aromatic, and Pesticidal Plants (MOAPPs) and the creation of jobs in the agro-industrial sector.

Medicinal plants may therefore be able to play a role in providing sustainable livelihood for the herbalists and in conserving valuable habitats and plant species. In addition, the crude plant extracts can be developed into useful drugs, and the chemicals extracted can be used to develop synthetic drugs or used as starting materials for the synthesis of new drugs in collaboration with scientists outside the country. A UNIDO consultancy on the industrial utilisation of medicinal and aromatic plants in Sierra Leone was conducted by Baser and MacFoy (1993); a pilot project was recommended but was not followed up by the country.

No doubt, there is great need for the conservation of endangered medicinal plant species through the implementation of strong national policies and for the cultivation of medicinal plants; development of botanic gardens; establishment of comprehensive databases on existing medicinal plants; and translation of relevant indigenous knowledge on the practice in local languages to enhance public awareness.

Use of sacred groves in medicinal plants conservation has already been suggested by Lebbie and Guries (1995). Sacred groves are forest patches conserved by the local people intertwined with their socio-cultural and religious practices. These groves harbor rich biodiversity and play a significant role in conservation. Indigenous cultural and ritual practices of the local people in sacred groves serve as a tool for conserving biodiversity.

Sacred groves in the Moyamba district of Sierra Leone were assessed for their value to local herbalists and traditional folk medicine practitioners of the Kpaa Mende people. Herbalists (*tufablaa*) collecting from twenty-three sacred groves were interviewed regarding their general knowledge of medicinal plants and their perceptions regarding changes in the occurrence of medicinal plants. Over seventy-five medicinal plant remedies are currently in use among the Kpaa Mende, with some plants reported here for the first time in terms of their medicinal uses in Sierra Leone. A significant feature of Kpaa Mende ethnobotany is the employment of two or more kinds of plants in combination as a remedy for a particular affliction. The discovery of rare and uncommon plants in the groves and their medicinal uses are discussed in terms of the role of the sacred groves in medicinal plant conservation (Lebbie and Guries 1995).

In Situ and Ex Situ Conservation

In situ conservation is regarded as the best means to ensure that the populations of species of plants and animals continue to grow and evolve in the wild, in their natural habitats. Conservation is achieved both by setting aside areas as nature reserves and national parks (collectively termed protected areas) and by ensuring that as many wild species as possible can continue to survive in managed habitats, such as farms and plantation forests.

Plant species should also be conserved ex situ (i.e., outside their habitat). The primary purpose of this is as an insurance policy. But it also has the advantage that it is usually easier to supply plant material for propagation, for reintroduction, for agronomic improvement, for research, and for education purposes from ex situ collections than from in situ reserves.

In Sierra Leone, a unique partnership was formed between a traditional healers' association, an academic research institute, and local communities; this cooperative should help to protect biodiversity and provide sustainable livelihoods for local communities through the establishment of the Tiwai Island Health and Fitness Center—a facility to provide health services based on principles of West African ethno-medicine (EFA, 2009).

Tiwai Island Health Village intends to protect biodiversity and provide sustainable livelihoods for local communities through the establishment of an eco-tourism facility, which will also provide health services based on principles of West African ethno-medicine.

Although this project aims to make some effort at conservation, more importantly, the small Gola forest in the southeast of Sierra Leone must also of necessity be conserved, since it remains the only remnant of primary forest remaining in the country; it could be housing valuable resources not yet discovered. In 2007, these 75,000 hectares of land were declared a national park. This park is part of the Upper Guinea Forest, one of the Global 25 Biological Diversity Hotspots. The only other forest reserve is Outamba Kilimi National Park in the north, an area of southern Guinea savanna woodlands, and some remote mangrove swamps, which was made a national park in 1986.

Property Rights

Protection of intellectual property (IP) rights and indigenous knowledge is an important component to the utilization of medicinal plants.

Development of medicinal plants relies very heavily on the knowledge carried by indigenous peoples and rural societies. In recent times, this has raised concerns about equitable sharing of the benefits of such knowledge and the IP rights of these indigenous rural communities. Securing such rights and protections entail specific dilemmas and policy implications.

Moreover, the Organization for African Unity, for example, has regulations that forbid export of medicinal plants in commercial quantities without the explicit permission of the host government. Research cooperation is likewise supposed to be regulated and monitored, but except for a few countries, these regulations are totally ignored.

Recent Developments in Traditional Medicine in Sierra Leone

From a historical perspective, perhaps the first recorded evidence of a Sierra Leonean combining his scientific training as a doctor with traditional medicine is that of Dr. John Abayomi-Cole (1848-1943) who, after leaving the CMS Grammar School and Fourah Bay College in Freetown, became an ordained minister in the United States and studied medicine there. He was a Fellow of the Apothecaries (F.S.A.) in America and an affiliate member of the National Association of Medical Herbalists of Britain. For his successful combination of both orthodox and traditional medicine using herbs, leaves, and roots, he became very famous throughout Sierra Leone and Liberia. During the Spanish influenza of 1918, he invented a preparation, made up of lime, teabush, camphor, and sprit, which saved the lives of many. He was also popular for his invention of *ekpa,* an antidote for poison and stomach problems. His practice was more popular and lucrative than those of orthodox doctors who studied in the UK, to the extent that he became the medical advisor to the governor, Sir Leslie Probyn.

Currently in Sierra Leone, the cost of imported drugs is very high; their inaccessibility and unavailability has resulted in many people using TM because of its cultural acceptability, affordability, and accessibility. At one time, a herbalist for broken limbs was attached to the Catholic hospital in Panguma, Kenema district.

Many African countries have now developed traditional medicine polices and regulations. For example, in Sierra Leone, the Ministry of Health and Sanitation, in its strategic plan for 2010-2015, stated as one of its objectives "the development of a traditional medicine policy."

Some of the information described below, which was obtained from the Ministry of Health's (MOH) website in August 2012, clearly describes the country's interest and the strides being made in TM (Health-SL, 2012). In summary, it states that traditional medicine currently forms part of the primary health care system in Sierra Leone. The traditional medicine program run by the Ministry of Health and Sanitation, has constructed a training school at Makeni, established a healing center at Kono, and conducted workshops at regular intervals to promote co-operation between traditional medicine practitioners and orthodox doctors. Members of the program are also locating, collecting and researching plants throughout Sierra Leone used for medicine.

The program gets support from WHO, and a consultant from Ghana, helped with developing the national policy for traditional medicine in keeping with WHO's mandates.

SLENTHA, the Sierra Leone Traditional Healers Association, is yet to have a permanent location, and the profession is faced with quacks and charlatans. These pose a threat, as there is the tendency for them to misuse traditional medicine. However, there is a Bill and Code of Ethics which is awaiting approval from the Sierra Leone parliament and cabinet. Even though the national policy on TM was launched in 2007, it has not yet been ratified by the Sierra Leone parliament as of August 2013

Once approved, the program would be able to control traditional healers according to the law; for example, the herbal vendors that parade the streets, hawking herbs, some of whom are not trained herbalists but obtain these herbs from the herbalists to sell, without any knowledge of the use of most of them.

After the bill is passed, only those certified by the association will be allowed to practice TM. It is envisioned that traditional medical practitioners will work alongside conventional medical counterparts to ensure an effective health service delivery in the country.

The ministry is expected to put into practice all the provisions embedded in the policy, since traditional medicine is part of the country's cultural heritage and should be promoted to contribute to the universal goal of affordable, accessible, appropriate, effective, and efficient health care services for all.

The policy would guide the ministry as they venture into the adoption and development of the ancient practice of traditional medicine. It outlines the key strategies that would ensure the development and appropriate utilisation and regulation of traditional medicines in Sierra Leone. This includes the preservation and protection of the flora and fauna with the potential for use as medicines.

The policy stipulates the development and enforcement of a Traditional Medicine Act and Code of Ethics and Standards of Practice. It also offers all stakeholders, including public, private, international, and nongovernmental organizations, the opportunity to collaborate with the government of Sierra Leone to advance traditional medicine practice in the country for the benefit of all (Health-SL,2012)

Traditional medicine is not systematically structured and regulated in the country to allow the population to reap its full benefits. Preparations have also not been subjected to assessment of quality, safety, efficacy, and affordability to guarantee rational use of the preparations.

Issues of conservation, cultivation, and preservation of medicinal plants of the country need to be considered along with the protection of the intellectual property rights and the benefits that may arise from the use of traditional medicine knowledge.

The country itself has some of the natural ingredients that constitute a good research atmosphere, but it is lacking in equipment and qualified personnel. However once this policy is implemented, it is anticipated that it would help enhance the development and appropriate use of traditional medicine in the health care system.

Prior to this new development in Sierra Leone, the regulatory situation of TM (i.e., the legal status), according to WHO, as of 2001, was as follows (WHO, 2001)

The website refers to the Medical and Dental Surgeons Act of 1966, which states that nothing in the act should be construed as prohibiting or preventing the practice of "customary systems of therapeutics," provided that such systems are not dangerous to life or health. The Medical Practitioners and Dental Surgeons Decree of 1994 repeals the Medical and Dental Surgeons Act of 1966. However, it retains exemptions for traditional medical practitioners.

The Traditional Medicine Act of 1996 regulates the profession of traditional medicine and controls the supply, manufacture, storage, and transportation of herbal medicines. It established a Scientific and Technical Board of Traditional Medicine and two committees under it: the Disciplinary Committee, which advises the board on matters relating to the professional conduct of traditional medicine practitioners, and the Drugs Committee, which advises the board on the classification and standardisation of traditional medicines.

The Scientific and Technical Board is charged with making sure that the highest practicable standards in the provision of traditional medicine in Sierra Leone are maintained by promoting the proper training and examination of students of traditional medicine, controlling the registration of traditional health practitioners, and regulating the premises where traditional medicine is practiced.

The Act further stipulates that the Board compiles a Register of Traditional Medical Practitioners, who then become members of the Sierra Leone Traditional Healers Association. Cancellation and suspension of registration, annual publication of the list of registered traditional medicine practitioners, restriction on use of the title Traditional Medical Practitioner, and the provision of medical aid by traditional medicine practitioners are also covered by the law. In addition, part four of the Act contains a list of the diseases for which traditional medical providers may not advertise treatments.

PART II.
Plants Species Grouped under Medicinal Uses in Alphabetical Order

1. Abnormal heart beat (palpitation; heart arrhythmia)
Cassia occidentalis
Microglossa pyrifolia

2. Abortifacient
Ficus exasperata
Peperomia pellucida
Portulaca oleracea
Solanum nodiflorum

3. Aid teething in babies
Erythrococca anomala

4. Anemia
Bidens pilosa
Ficus capensis

5. Antidote
Vismia guineensis

6. Aphrodisiac
Striga macrantha

7. Appetite
Agelacea oblique

Pseudospondias microcarpa

8. Arthritis
Vitellaria paradoxa/Butyrospermum parkii

9. Asthma
Corchorus olitorus
Desmodium adscendens
Pennisetum purpureum
Rottboellia exaltata

10. Avert bedwetting
Mezoneurum benthamianum

11. Bad breath
Strychnos afzelia

12. Bleaching agent
Costus afer

13. Bilharzia
Anchomane difformis
Macaranga barteri
Secamone afzelii
Smilax kraussiana

14. Boils
Chasmanthera dependens
Datura metel
Elaeis guineensis
Euphorbia deightonii
Lycopersicum esculentum
Parinari excelsa
Selaginella myosurus
Sida stipulate
Tamarindus indica

15. Cause disease in thieves
Gossypium hirsutum

16. Closure of frontal suture
Anthocleista nobilis
Anthonotha macrophylla
Ficus capensis
Hibiscus esculentus

17. Common cold
Allophylus africanus
Clematis grandifolia

18. Constipation
Alchornea cordifolia
Carica papaya
Cassia siberiana

Ficus umbellata
Momordica charantia
Tamarindus indica

19. Contraceptive
Ananas comosus
Sida stipulata

20. Convulsion
Albizia adianthifolia
Bryophyllum Pinnatum
Ocimum basilicum
Cassia occidentalis
Cnestis ferruginea
Ficus exasperata

Mussaenda erythrophylla
Ocimum americanum
Ocimum gratissimum
Ocimum viride
Selaginella myosurus

21. Cough
Abrus precatorius
Aframomum melegueta
Alchornea cordifolia
Amaralia heinsioides
Ananas comosus
Cajanus cajan
Cassia occidentalis
Craterispermum laurinum
Dialium guineense

Dissotis rotundifolia
Garcinia kola
Hibiscus surattensis
Manniophyton fulvum
Palisota hirsuta
Salacia senegalensis
Tetracera leicarpa
Trema guineensis
Trichilia heudelotii

22. Development of premature children
Alchornea cordifolia

23. Diabetes
Cassia occidentalis

24. Diseases of the ear
Bryophyllum pinnatum
Palisota hirsuta

25. Diseases of the eyes
Amaralia heinsioides
Cola nitida
Dissotis rotundifolia

Manihot esculenta
Sarcocephalus esculentus

26. Diseases of the gum
Cnestis ferruginea

27. Drunkenness
Annona muricata

28. Dysentery (diarrhea)
Albizia zygia
Amphimas pterocarpoides
Phyllanthus discoideus
Anchomanes difformis
Diospyros thomasii
Gossypium hirsutum
Mammea americana
Manniophyton fulvum

Ocimum gratissimum
Parinari excelsa
Phyllanthus muellerianus
Psidium guajava
Sesamum indicum
Tetracera alnifolia
Tetracera potatoria

29. Ease labor (at childbirth)
Ancistrophyllum secundiflorum
Desmodium adscendens
Dioscorea bulbifera
Spondias mombin

30. Elephantiasis
Newbouldia laevis

31. Enlightens slow children
Byrsocarpus coccineus

32. Epilepsy
Eupatorium sp.

33. Evil spirit
Allium cepa
Capsicum frutescens
Mimosa pudica

34. Fertility
Dioscorea bulbifera
Peperomia pellucida

35. Fetal development
Gouania longipetala

36. Fever
Albizia adianthifolia
Allophylus africanus
Ananas comosus
Bryophyllum pinnatum
Capsicum frutescens
Lippia adoensis
Cassia occidentalis
Chassalia kolly
Chassalia kolly
Ocimum basilicum
Citrus aurantifolia

Cymbopogon citratus
Eryngium foetidum
Eupatorium sp.
Ficus umbellate
Hyptis suavolen
Microglossa afzelli
Morinda geminata
Nauclea latifolia
Strychnos afzelia

37. Fish poison
Adenia lobata
Raphia vinifera

Stephania dinklagei
Tephrosia vogelii

38. Fragrance and aphrodisiac
Strychnos afzelia

39. Frequent urination
Dissotis rotundifolia

40. Gonorrhea

Ageratum conyzoides
Citrus aurantifolia
Costus afer
Craterispermum laurinum
Dialium guineense
Dissotis rotundifolia
Harungana madagascariensis
Lophira lanceolata
Macaranga heterophylla
Nauclea latifolia
Paullinia pinnata
Phyllanthus muellerianus
Scoparia dulcis
Secamone afzelii
Tetracera leicarpa

41. Hardened breasts in breastfeeding mothers

Mangifera indica

42. Headache

Afzelia sp (not Africana)
Allophylus africanus
Carica papaya
Citrus aurantifolia
Cinnamomum zeylanicum
Dioscorea bulbifera
Ficus umbellata
Mimosa pudica
Musa paradisiaca
Ricinus cummunis
Scoparia dulcis
Spondias mombin

43. Healing of navel stump in newborns

Bryophyllum pinnatum

44. Heal enlarged spleen

Portulaca oleracea

45. Healing after female circumcision

Nauclea latifolia

46. Heart attack

Anchomanes difformis

47. Heart disease

Alchornea cordifolia
Nauclea latifolia

Microglossa pyrifolia

48. Hemorrhoids

Combretum sp.
Dichrostachys glomerata
Phyllantus discoideus

Solanum nodiflorum
Tetracera potatoria

49. Hemostasis

Alchornea cordifolia
Amphimas pterocarpoides
Ipomoea involucrate

Nicotiana sp.
Portulaca oleracea
Rutidea olenotricha

50. Hernia

Citrus aurantifolia
Clerodendrum umbellatum

Parinari excelsa
Portulaca oleracea

51. Hiccups

Diospyros hendelotii
Elaeis guineensis
Microdesmis puberula

Musa sapientum
Tamarindus indica
Zea mays

52. Husky voice in newborns

Spondias mombin

53. Hypertension

Cassia occidentalis

54. Improvement of milk quality

Anisophyllea laurina

55. Increase libido/aphrodisiac

Garcinia kola
Stringa macrantha

56. Increase strength in children

Alternanthera sessilis
Chlorophora regia
Dracaena surculesa

Microglossa pyrifolia
Setaria chevalieri

57. Induce labor
Microdesmis puberula

58. Induce sleep
Mussaenda erythrophylla

59. Induce vomiting when poisoned
Limnophyton obtusifolium

60. Inflammation

Calyptrochilum emarginatum
Datura metel
Elaeis guineensis
Euphorbia deightonii

Lycopersicum esculentum
Mimosa pudica
Secamone afzelii
Sida stipulata

61. Insecticide
Ageratum conyzoides
Oldfieldia africana

62. Itch
Ficus exasperata
Microdesmis puberula

63. Jaundice
Polyscias sp.

64. Laxative

Anthocleista nobilis
Carapa procera
Cassia alata
Cassia siberiana
Mareya micrantha
Morinda geminata
Nauclea latifolia

Ocimum viride
Sarcocephalus esculentus
Selaginella myosurus
Smeathmannia pubescens
Spondias mombin
Uvaria chamae

65. Lice
Cassia podocarpa

66. Malaria

Alchornea cordifolia

Azadirachta indica
Carapa procera
Cassia siamea
Cassia siberiana
Craterispermum laurinum
Erythrococca anomala
Ficus umbellata
Harungana madagascariensis

Lantana camara
Mareya micrantha
Momordica charantia
Morinda geminata
Morinda morindoides
Nauclea latifolia
Persea americana
Phyllanthus muellerianus
Spondias mombin
Uvaria chamae

67. Mosquito repellant

Hyptis suavolens

68. Measles

Aframomum elliotii
Cajanus cajan

Mareya micrantha
Spondias mombin

69. Menstrual pain

Ocimum Basilicum

70. Nausea

Microdesmis puberula

71. Pain

Ageratum conyzoides
Anthocleista vogelii
Caloncoba brevipes
Cassia siberiana
Cleistopholis patens
Clerodendrum umbellatum
Cyanotis lanata
Desmodium adscendens
Dichrostachys glomerata
Dioscoreophyllum spp.
Dracaena arborea

Newbouldia laevis
Ocimum americanum
Ocimum viride
Ouratea morsonii
Parinari excelsa
Rauwolfia vomitoria
Secamone afzelii
Spondias mombin

Tapinanthus banguensis
Triclisia patens

Mareya micrantha
Mimosa pudica

Vitex doniana

72. Piles
Ocimum bacilicum

73. Potency
Carpodinus dulcis
Fagara macrophylla

Landolphia dulcis
Mezoneurum
benthamianum

Hibiscus sabdariffa

Mimosa pudica

74. Pregnancy detection
Cola nitida
Psychotria rufipilis

75. Prolonged menstruation
Alchornea cordifolia
Canthium glabriflorum
Sabicea vogelii

76. Protection against witches
Eryngium foetidum

77. Rash
Ficus exasperata

78. Rat poison
Dichapetalum toxicarium
Erythrophylum guineense

79. Rheumatism
Butyrospermum parkii
Carapa procera
Erythrophleum guineense
Carica papaya
Premna hispida
Uvaria chamae

80. Scorpion bite
Mammea americana

81. Skin diseases
Ageratum conyzoides
Amaranthus hybridus
Borreria verticallata
Cassia alata
Desmodium adscendens
Diospyros heudelotii
Dracaena sp.
Erigeron floribundus
Ficus exasperata

Mangifera indica
Mareya micrantha
Mimosa pudica
Microdesmis puberula
Newbouldia laevis
Peperomia pellucida
Rauvolfia vomitoria
Selaginella myosurus
Scleria boivinii

82. Snake bite
Desmodium adscendens
Dracaena arborea

83. Snake repellent
Datura metel
Dichrostachys glomerata

84. Sores
Bidens pilosa
Carica papaya
Jatropha curcas
Manihot esculenta

Solanum nodiflorum
Sterculia tragacantha
Vitex doniana

85. Sore throat
Aframomum melegueta
Alchornea cordifolia

Craterispermum laurinum
Pennisetum purpuereum

86. Splitting feet
Ficus capensis

87. Stomachache
Carapa procera
Gongronema latifolium

Strychnos afzelia

88. Stomach poison
Aframomum melegueta
Harungana madagascariensis

Musa sapientum
Secamone afzelii

89. Stop vomiting
Capsicum frutescens
Citrus aurantifolia
Diospyros heudelotii

Ocimum americanum
Premna hispida
Triumfetta cordifolia

90. Strengthen body
Adansonia digitata
Harungana madagascariensis
Newbouldia laevis

Uvaria chamae
Zea mays

91. Strength/increase weight
Adansonia digitata

92. Success in life
Allophylus gracilis

93. Swelling
Alchornia cordifolia
Cyanotis lanata

Eryngium foetidum
Ricinus communis

94. Swollen stomach
Cercestis afzelia

95. Thirst
Ananas comosus

96. Thorn prick
Tetracera alnifolia

97. Tonic
Allophylus africanus
Canthium glabriflorum
Harungana madagascariensis

Myranthius serratus
Vismia guineensis

98. Toothache
Acacia pennata
Anacardium occidentale

Cajanus cajan
Caloncoba echinata
Carica papaya
Cola caricaefolia
Elaeis guineensis

Harungana
madagascariensis
Microglossa pyrifolia
Musanga cecropicoides
Newbouldia laevis
Phyllanthus discoideus
Uncaria africana

99. Tuberculosis
Tetracera potatoria

100. Urinary tract blockage in men
Ageratum conyzoides

101. Urinary tract infection
Tetracera leicarpa

102. Vaginal infection
Nauclea latifolia
Ocimum viride

103. Worms
Anchomane difformis
Aspilia latifolia
Bersama abyssinica
Butyrospernum parkii
Carica papaya
Dichrostachys glomerata
Ficus exasperate
Harungana madagascariensis

Newbouldia laevis
Ocimum gratissimum
Phasedes lunatus
Piper umbellatum
Scoparia dulcis
Urera rigida
Vernonia amygdalina

104. Wounds
Ageratum conyzoides
Alchornea cordifolia
Parinari excelsa
Alternanthera sessilis

Aspilia latifolia
Leptoderris trifoliate
Pennisetum subangustrum
Tamarindus indica

105. Yellow fever

Ageratum conyzoides
Anthonotha macrophylla
Carica papaya
Costus afer

Craterispermum laurinum
Jatropha curcas
Leonotis nepetifolia

PART III.

Plant Family (in Alphabetical Order), Species, Uses, and Method of Drug Preparation

In some cases, both the old and the new family names are used for clarity purposes.

The letters used refer to the area from which the plants were collected (G = Gloucester Village, P = Pujehun district; H = Hastings Village; Ko = Kono district; PL = Port Loko district; Ma = Makeni district; Ke = Kenema district; F= Freetown) or the local name of the plant (M = Mende; K = Krio; T = Themne; S = Soosoo; L = Loko; Ko = Kono).

Family: *AGAVACEAE*
Botanical Name: *Dracaena sp.*   **Local Name**: Baboo kola (K)
Medicinal Use: Skin disease "allay" (K) (H)
Preparation and Treatment: The roots are dug up at dawn or dusk; then roasted together with the roots of *Carapa procera*. Next palm oil is mixed with the ashes from a fireplace to form a poultice which is then applied to the skin, three times a day.

Family: *AGAVACEAE*
Botanical Name: *Dracaena arborea* Link   **Local Name:** Kokhat (K)
Medicinal Use (1): Snake bite (G)
Preparation and Treatment (1): After the young leaves are removed from the rosette, the pale white lower portion embedded in the rosette is rubbed on the affected area.
Medicinal Use (2): Ease period pains with heavy clotting (H)

Preparation and Treatment (2): A decoction of *Dracaena arborea* leaves together with *Newbouldia* and *Nauclea Latifloia* leaves is taken orally.

Family: *AGAVACEAE*
Botanical Name: *Dracaena surculosa* Lindl.   **Local Name:** Wete (Ko)
Medicinal Use: Increase strength and physical fitness in children (Ko)
Preparation and Treatment: The leaves are boiled, and the decoction used to bathe the child twice a day. About 50 ml can also be drunk concurrently.

Family: *ALISMATACEAE*
Botanical Name: *Limnophyton obtusifolium*   **Local Name:**
Miq.   Tagbo-Mese (Ko)
Medicinal Use: Induce vomiting when poisoned (Ko)
Preparation and Treatment: The bulb is crushed and boiled, and the decoction taken immediately after poisoning, sometimes with palm oil.

Family: *ALLIACEAE*
Botanical Name: *Allium cepa* Linn.   **Local Name:** Yabas (K)
Medicinal Use: Repel evil spirits (H)
Preparation and Treatment: Onions, garlic, and spice (*Xylopia*) are pounded together and then mixed with water and some Florida water (a perfume); this is then applied on the child's body in the morning and evening.

Family: *AMARANTHACEAE*
Botanical Name: *Alternanthera sessilis*   **Local Name:**
(Linn.) B. Br. ex Schult   nDatawuli (M)
Medicinal Use (1): Wounds (P)
Preparation and Treatment (1): A decoction of the leaves is used to wash the wound or the juice from the leaves is squeezed on to the wound.
Medicinal Use (2): Strengthens children (P)(Ke)
Preparation and Treatment (2): A decoction of the leaves is used to bathe children to give them strength (P). A decoction of the leaves together with the stem is used to bathe the child; a handful of the decoction is also given orally before every bath (Ke).

Family: *AMARANTHACEAE*
Botanical Name: *Amaranthus hybridus* Linn.   **Local Name**: Greens
(Figure 5*)*   (wild variety) (K)
Medicinal Use: Skin disease "allay" (K)
Preparation and Treatment: Leaves together with the leaves of *Passiflora quadrangularis* are macerated with palm oil and applied to the affected area three times a day.

Figure 5—*Amaranthus hybridus*

Family: *ANACARDIACEAE*
Botanical Name*: Anacardium occidentale* Linn.   **Local Name**: Kushu (K)
Medicinal Use: Toothache and sore gums (F)
Preparation and Treatment: An infusion of the leaves and bark is used as a mouthwash:

Family*: ANACARDIACEAE*
Botanical Name: *Mangifera indica* Linn. (Figure 6)   **Local Name:** Mangro (K)
Preparation and Treatment:
Medicinal Use (1): Scabies and other skin diseases (G)
Preparation and Treatment (1): A decoction of the bark is applied to the affected area for at least three days before removal.
Medicinal Use (2): Hardened breasts in breastfeeding mothers (H)

Preparation and Treatment (2): The bark of the plant is pounded with a pestle and mortar and mixed with water, and then filtered. The filtrate is made into a paste with ash from the fireplace and applied to the breast.

Figure 6—*Mangifera indica*

Family: *ANACARDIACEAE*
Botanical Name: *Pseudospondias microcarpa* (A. Rich.) Engl. **Local Name:** Dueke (Ko)
Medicinal Use: Increase appetite (Ko)
Preparation and Treatment: The bark of the tree together with the young leaves are boiled, and the decoction is drunk four times a day.

Family: *ANACARDIACEAE*
Botanical Name: *Spondias mombin Linn.* (Figure 7) **Local Name:** Fix plum (K)
Local name: gBojie (M)
Local name: Kpangine (Ko)
Medicinal Use (1): Ease labor (P) (Ko)
Preparation and Treatment (1): A decoction of the leaves is drunk.

Medicinal Use (2): Headache (G)

Preparation and Treatment (2): Leaves are gently heated over a flame and tied to the head.

Medicinal Use (3): Husky voice in newborn babies (H)

Preparation and Treatment (3): A decoction of *Spondias mombin* leaves is drunk (~20 ml) in the morning.

Medicinal Use (4): Laxative (P)

Preparation and Treatment (4): A decoction of the leaves is drunk.

Medicinal Use (5): Malaria (G)

Preparation and Treatment (5): An infusion of the leaves with those of *Carica papaya* and *Morinda geminata* is used for bathing.

Medicinal Use (6): Measles (G)

Preparation and Treatment (6): An infusion of the leaves is drunk.

Medicinal Use (7): Stomach pains after childbirth (P)

Preparation and Treatment (7): A decoction of the leaves is drunk.

Figure 7—*Spondias mombin*

Family: *ANNONACEAE*

Botanical Name: *Annona muricata* Linn.   **Local Name:** Sawa Sap (K)

Medicinal Use: Relieve drunkenness (H)

Preparation and Treatment: Leaves are macerated with some water in a calabash and then rubbed on the patient's face continuously until sobriety is achieved.

Family: *ANNONACEAE*
Botanical Name: *Cleistopholis patens* (Benth.) **Local Name:**
Engl. & Diels Moigbenei (M)
Medicinal Use: Joint pain (P)
Preparation and Treatment: A soft poultice obtained by crushing young leaves in a small quantity of water is applied to the joints, especially to the wrist.

Family: *ANNONACEAE*
Botanical Name: *Uvaria chamae P. Beauv.* **Local Name:** Finga (K)
(Figure 8)
Medicinal Use (1): Laxative (G)
Preparation and Treatment (1): A decoction of the roots with the dried leaves of *Nauclea latifolia* is drunk.
Medicinal Use (2): Malaria (G)
Preparation and Treatment (2): An infusion or a decoction of the dried leaves with those of *Sterculia tragacantha* is used to bathe.
Medicinal Use (3): Rheumatism (G)
Preparation and Treatment (3): A decoction of the root is drunk.
Medicinal Use (4): Strengthen the body (H)
Preparation and Treatment (4): A decoction of the *Uvaria chamae* leaves together with the leaves of *Nauclea latifolia* and *Morinda geminata* is used to bathe, twice a day.

Figure 8—*Uvaria chamae*

Family: *ANNONACEAE*
Botanical Name: *Xylopia aethiopica*
(Dunal) A. Rich. (Figure 9)

Local Name: Hiwii (M),
Local Name: Spice-tick
(K)

Medicinal Use: Cough (Ke)
Preparation and Treatment: A decoction of the mature dried fruits is drunk. Alternately, leaves can be chewed and the juice swallowed.

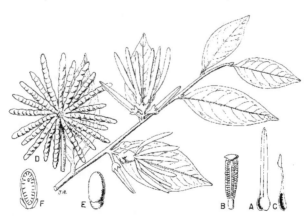

Xylopia aethiopica (Dunal) A. Rich. (*Annonaceae*). **A**, petal. **B**, stamen. **C**, carpel.
), fruits. **E**, seed with aril. **F**, vertical section of seed (orig.).

Figure 9—*Xylopia aethiopica*

Family: *APOCYNACEAE*
Botanical Name: *Carpodinus dulcis*   **Local Name**:
(R. Br.) Pichon   Kushument (K)
Medicinal Use: Impotence (H)
Preparation and Treatment: The roots are soaked in a bowl of water for four hours and then boiled, and the decoction obtained is drunk as the drug.

Family: *APOCYNACEAE*
Botanical Name: *Landolphia dulcis*   **Local Name:**
(R. Br.) Pichon   Kushument (K)
Medicinal Uses: Potency (G)
Preparation and Treatment: The bark of the tree is chewed and the juice swallowed or a maceration of the bark is drunk.

Family: *APOCYNACEAE*
Botanical Name: *Rauvolfia vomitoria* Afzel.   **Local Name:** Kowogei
                                                    (M)
                                                    Local Name: Yokpa (T)

Medicinal Use (1): Pain (G)
Preparation and Treatment (1): A decoction of the leaves is drunk and the remainder and subsequent extracts used to bathe.
Medicinal Use (2): Scabies (Ke)
Preparation and Treatment (2): The young leaves are wrapped in a *Mitragyna stipulosa* leaf and tied, and heated in hot wood ash. The young leaves are then squeezed and the liquid mixed with white clay and applied as a lotion.

Family: *ARACEAE*
Botanical Name: *Anchomanes difformis*   **Local Name:** Kiponi
(Bl.) Engl.   (M)
                                               Local Name: Pondi
                                               (Ko)
                                               Local Name: a-tho.
                                               bothigba (T)
Medicinal Use (1): Diarrhea (Ma)
Preparation and Treatment (1): A leaf decoction is drunk.

Medicinal Use (2): Heart attack (Ke)

Preparation and Treatment (2): The stem is dried and burned to ashes. At time of attack, the ash dissolved in palm oil is administered orally (dosage: the equivalent of four fingertips for men and three for women, four times a day).

Medicinal Use (3): Kill worms, especially *Schistosoma haematobium* (Bilhazia worms) (Ko)

Preparation and Treatment (3): The rhizome of the plant with the outer skin removed is boiled, and the decoction drunk daily.

Family: *ARACEAE*

Botanical Name: *Cercestis afzelia* Schott   **Local Name:** mBembei (M)

Medicinal Use: Swollen stomach (P) (called "GoBoi" in Mende)

Preparation and Treatment: The filtrate obtained from pounded leaves and water is boiled and the decoction is drunk.

Family: *ARALIACEAE*

Botanical Name: *Polyscias* sp.   **Local Name:** Snow White (K)

Medicinal Use: Jaundice (G)

Preparation and Treatment: An infusion of a small quantity of the leaves together with those of *Momordica charantia* and *Morinda geminata* is drunk.

Family: *ASCLEPIADACEAE*

Botanical Name: *Gongronema latifolium* Benth.   **Local Name:** Tonenyuu (Ko)

Medicinal Use: Stomachaches (Ko)

Preparation and Treatment: The leaves are pounded and water and common salt (sodium chloride) added; the filtrate is taken twice a day.

Family: *ASCLEPIADACEAE*

Botanical Name: *Secamone afzelii* (Schultes) K. Schum.   **Local Name:** Boaneh (Ko)

Local Name: nDikpagbaa (M)

Medicinal Use (1): Bilharzia (P)

Preparation and Treatment (1): Pounded leaves are made into a sauce and eaten with rice.

Medicinal Use (2): Gonorrhea (P)

Preparation and Treatment (2): Pounded leaves are made into a sauce and eaten with rice.

Medicinal Use (3): Inflammation (P)

Preparation and Treatment (3): Pounded leaves are applied as a poultice to the affected areas.

Medicinal Use (4): Stomach pains (Ko)

Preparation and Treatment (4): The leaves are boiled and the decoction is drunk; approximately 150 ml for adults and 100 ml for children.

Medicinal Use (5): Stomach poison (caused by witches) (P)

Preparation and Treatment (5): Pounded leaves are made into a sauce and eaten with rice.

Family: *BIGNONIACEAE*

Botanical Name: *Newbouldia laevis* Seem.   **Local Name:** Snuff leaf (K)

Medicinal Use (1): Skin disease; "allay" (K) (H)

Preparation and Treatment (1): The roots are chopped into small pieces and then roasted in a pot containing spice (*Aframomum melegueta*) to obtain a thick black residue. The residue is made into a poultice using palm oil and nut oil, and applied to the affected skin, twice a day.

Medicinal Use (2): Stomach pains (H)

Preparation and Treatment (2): A decoction of small pieces of the roots and spice (*Xylopia*) is drunk.

Medicinal Use (3): Strengthen one's body (H)

Preparation and Treatment (3): A decoction of the leaves of *Newbouldia* together with *Nauclea latifolia* and *Ficus capensis* is used to bathe, two times a day.

Medicinal Use (4): Toothache (H)

Preparation and Treatment (4): A small piece of the root is placed in the tooth cavity.

Medicinal Use (5): For treatment of worms (H)

Preparation and Treatment (5): A decoction of the bark of the *Newbouldia* plant is drunk.

Medicinal Use (6): Elephantiasis (H)

Preparation and Treatment (6): Pounded leaves and stem are heated and applied hot to the affected area.

Family: *BOMBACACEAE*

Botanical Name: *Adansonia digitata* Linn. (Figure 10)

Local Name: Baobab or monkey bread (K)

Local Name: An-der-a-bai (T)

Medicinal Use (1): Strengthens the body (H)
Preparation and Treatment (1): A decoction of the bark of the plant is used to bathe (not to be used on the head).
Medicinal Use (2): Strength/increased weight (Ma)
Preparation and Treatment (2): A root decoction mixed with food is eaten.

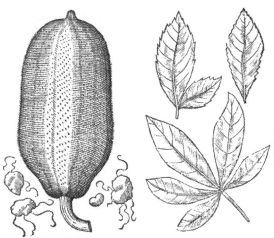

Figure 10—*Adansonia digitata*

Family: *BROMELIACEAE*

Botanical Name: *Ananas comosus* (Linn.) Merrill

Local Name: Pineapul or pineapple (K)

Local Name: A-nanas (T)

Medicinal Use (1): Cough (G)
Preparation and Treatment (1): An infusion of the leaves is drunk.

Medicinal Use (2): Contraceptive (H)
Preparation and Treatment (2): The leaves are pounded and twisted together with the leaves of *Gossypium* into a cord; the user then expresses her wishes in secret, holding the cord in both hands.
Medicinal Use (3): Thirst (Ma)
Preparation and Treatment (3): The fruit is used to quench thirst (especially when one has a fever).

Family: *CARICACEAE*
Botanical Name: *Carica papaya* Linn.   **Local Name:** Fakai (M)
                                              Local Name: Kongbi (Ko)
                                              Local Name: Pawpaw (K)

Medicinal Use (1): Headache (H)
Preparation and Treatment (1): A poultice is made with the papaya seeds and nut oil and then applied to the shaved head.
Medicinal Use (2): Rheumatism (H)
Preparation and Treatment (2): The roots are pounded and placed in a white cloth, and the cloth tied around the painful areas (drug generates a lot of heat).
Medicinal Use (3): Toothache (G) (H)
Preparation and Treatment (3): The epidermis is removed from the stem, and the remaining bark is ground with salt and inserted into the cavity (G). A small piece of the root is inserted in the cavity (H).
Medicinal Use (4): Worms (Ma)
Preparation and Treatment (4): The seeds are eaten to expel worms.
Medicinal Use (5): Yellow fever and constipation (Ko)
Preparation and Treatment (5): The leaves are dried in the sun for a day and then boiled. The decoction is drunk three times a day; ~150 ml for adults and ~100 ml for children.
Medicinal Use (6): Yellow fever (P)
Preparation and Treatment (6): A decoction of the dried leaves is drunk.
Medicinal Use (7): Sores (F)(H)
Preparation and Treatment (7): The rind or skin of the pawpaw fruit is placed on the sore (F); a mixture of the crushed leaves and pounded roots is applied to the sore.(H).

Family: *CELASTRACEAE*
Botanical Name: *Salacia senegalensis* (Lam.) DC

Local Name: Gigboi (M)
Local Name: Malombo (K)

Medicinal Use: Cough (P)
Preparation and Treatment: For children, young leaves are wrapped in larger ones and heated. The juice from the young leaves is squeezed out and taken orally as a drug. Adults can chew the leaves and swallow the juice.

Family: *CHRYSOBALANACEAE*
Botanical Name: *Parinari excelsa* Sab.

Local Name: N' Dawei (M)
Local Name: Rof-skin-plom (K),

Medicinal Use (1): Stomach pain in males (H)
Preparation and Treatment (1): A decoction of the bark from the plant is drunk.
Medicinal Use (2): Stomach pain (G)
Preparation and Treatment (2): A decoction of the leaves with fruits of *Xylopia* (spice) is drunk.
Medicinal Use (3): Boils (hard, poisoned swelling under the skin, which bursts when ripe) (G)
Preparation and Treatment (3): The bark is pounded and applied to the boil.
Medicinal Use (4): Hernia (Ke)
Preparation and Treatment (4): The bark of mature stem is pounded into a powder and then mixed with white clay. A colloidal suspension obtained by dissolving a fraction of this solid dried paste in water is filtered and the filtrate drunk.

Family: *COMBRETACEAE*
Botanical Name: *Combretum sp.*

Local Name: Fly lif (K)

Medicinal Use: Hemorrhoids (G)
Preparation and Treatment: The juice obtained from the leaves is drunk.

Family: *COMMELINACEAE*

Botanical Name: *Cyanotis lanata* Benth. **Local Name:** Fibrous grass (K)

Medicinal Use: Swollen scrotum and body pain (H)

Preparation and Treatment: Finely cut leaves of *Cyanotis* are pounded together with the leaves of *Portulaca oleracea* and then sun dried. The dried leaves are then pounded again, sieved, and mixed with *Aframomum melegueta* (alligata pepe) and camphor; the powder obtained is then mixed with water to form a thick paste and then applied to the swelling.

Family: *COMMELINACEAE*

Botanical Name: *Palisota hirsuta* (Thunb.) K. Schum. **Local Name:** nDomu (M)

Local Name: Tugbe (Ko)

Medicinal Use (1): Cough (G)

Preparation and Treatment (1): The stem is chewed.

Medicinal Use (2): To stop pus production from ear due to infection (Ko)

Preparation and Treatment (2): The stem is heated over fire and the outer epidermis removed, and the remaining stem is then pounded with the seeds of a spice called "*Nyamoe*" in Kono (i.e. *Aframomum melegueta*) and a few drops of water. The liquid obtained on squeezing this pounded mixrure is applied into the infected ear (two drops, three times and a day).

Family: *COMPOSITAE (ASTERACEAE)*

Botanical Name: *Ageratum conyzoides* Linn. **Local Name**: Wait Aid Leaf (K),

Local Name: Watered grass (K)

Local Name: Yarnigbei (M)

Local Name: Yani-gbe (M)

Local Name: Yarndigbe (Ko)

Medicinal Use (1): Gonorrhea (H)

Preparation and Treatment (1): The leaves together with spice (*Xylopia*) are pounded in a mortar and the contents emptied into an enamel bowl with water. The liquid obtained is drunk while standing at a door entrance (~100 ml, twice a day).

Medicinal Use (2): Inactivate stinging insects, particularly wasps (Ke)

Preparation and Treatment (2): Leaves are placed near the insect.

Medicinal Use (3): Lower abdominal pain (H)

Preparation and Treatment (3): After pounding the leaves with spice (*Xylopia*) in a mortar, the mixture is squeezed between the palms; the liquid obtained is mixed with palm oil and drunk (~ 250 ml, one or two times a day)

Medicinal Use (4): Skin disease, Tinea (called *Facinclot* in Krio)(G)

Preparation and Treatment (4): Leaves are rubbed between the palms and the juice applied as a lotion to the affected area.

Medicinal Use (5): Urinary tract blockage in men (H)

Preparation and Treatment (5): A decoction of the leaves with leaves of *Portulaca oleracea* and spice (*Xylopia*) is taken orally (~50 ml, three to four times a day).

Medicinal Use (6): To cure light wounds (Ko), wounds (P)

Preparation and Treatment (6): The leaves are crushed between the palms and the liquid obtained is applied as the drug to a new laceration or wound. It may be reapplied if necessary (Ko). The juice obtained from the leaves is applied to light wounds (P).

Medicinal Use (7): Yellow fever (P)

Preparation and Treatment (7): A decoction of the leaves and stem together with the stem of *Craterispermum laurinum* (without epidermis) is drunk.

Family: *COMPOSITAE (ASTERACEAE)*

Botanical Name: *Aspilia latifolia* Oliv. & Hiern　　　　**Local Name:** Tonyei (M)

Medicinal Use (1): Wounds (Ke)

Preparation and Treatment (1): Fresh, cleaned leaves are crushed by rubbing them between the palms and the fluid squeezed into a clean container and applied to wounds.

Medicinal Use (2): Intestinal worms (Tapeworm exclusively)(Ke)

Preparation and Treatment (2): The leaves are macerated with a small amount of water, and a bit of common salt is added. The liquid obtained is drunk as the drug.

Figure 11—*Aspilia africana*

Family: *COMPOSITAE (ASTERACEAE)*
Botanical Name: *Bidens pilosa* Linn.
(Figure 12)

Local Name: Nidul lif
(K)
Local Name: Kandane
(Ko)

Medicinal Use (1): Anemia (P)
Preparation and Treatment (1): The juice obtained from crushed leaves is drunk early in the morning before meals.
Medicinal Use (2): Heals sores on penis or vagina due to *Trichomonas*. Locally called "m*arkru*" in Krio (Ko)
Preparation and Treatment (2): The leaves are burned and palm oil added to make a soft paste. This poultice is applied evenly on the sores and douched with hot water twice a day.

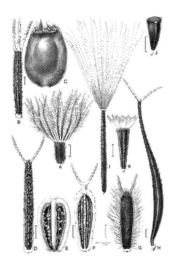

Figure 12—*Bidens pilosa*

Family: *COMPOSITAE (ASTERACEAE)*
Botanical Name: *Erigeron floribundus*   **Local Name:**
(H. B. & K.) Sch. Bip.   N'Jopor—yakpoi (M)
Medicinal Use: Parasitic skin disease, Tinea—"Facinclot" K) (Ke)
Preparation and Treatment: The leaves are macerated on a flat surface
with cold water and the poultice applied to the affected areas.

Family: *COMPOSITAE (ASTERACEAE)*
Botanical Name: *Eupatorium sp.*   **Local Name:**
                                                 Yamba-Dukue Musuma
                                                 (Ko)
Medicinal Use (1): Fever (Ko)
Preparation and Treatment (1): The stem and leaves are boiled and the
decoction obtained is drunk as the drug.
Medicinal Use (2): Epilepsy
Preparation and Treatment (2): The stem and leaves are boiled and the
decoction obtained is drunk as the drug.

Family: *COMPOSITAE (ASTERACEAE)*
Botanical Name: *Microglossa afzelli* O. Ktze.   **Local Name:**
                                                     Kie-keh-Bera (T)
Medicinal Use: Fever (G)

Preparation and Treatment: An infusion of the leaves is drunk or a poultice from the crushed leaves mixed with palm oil is applied as a liniment over the patient's body.

Family: *COMPOSITAE (ASTERACEAE)*
Botanical Name: *Microglossa pyrifolia*   **Local Name:**
(Lam.) O. Ktze.   Gimbo-yuvii (M)
Medicinal Use (1): Toothache (P)
Preparation and Treatment (1): A decoction of the leaves is used as a gargle.
Medicinal Use (2): Heart disease (P),
Preparation and Treatment (2): Leaves together with those of *Psychotria rupifilis* are pounded with water. White clay is added to the filtrate until a soft paste is obtained, which is then dissolved in water and drunk. Alternatively, the leaves of *Microglossa* are chewed with kola nuts (*Garcinia kola*)
Medicinal Use (3): Strengthens children (P)
Preparation and Treatment (3): A decoction of the leaves is used to bathe young children to give them strength.

Family: *COMPOSITAE (ASTERACEAE)*
Botanical Name: *Vernonia amygdalina* Del   **Local Name**: Bitter leaf
    (K)
Medicinal Use: Worms (H)
Preparation and Treatment: The leaves are pounded and the juice squeezed out and drunk or a decoction of the leaves is drunk (~150 ml, drunk once).

Family: *CONNARACEAE*
Botanical Name: *Agelaea obliqua*   **Local Name:** Weebuii
(P. Beauv.) Baill.   (Ko)
Medicinal Use: Gives appetite to anaemic patients (Ko)
Preparation and Treatment: Leaves are pounded and water added, and left to stand for some time. The filtrate obtained is used to prepare a daily meal.

Family: *CONNARACEAE*
Botanical Name: *Byrsocarpus coccineus*
Schum. & Thonn

Local Name: Yarin leaf (K)

Medicinal Use: Enlightens slow children (H)
Preparation and Treatment: The leaves together with groundnuts (*Arachis hypogea*) are pounded, dried, and mixed with fresh cow milk in a bottle; the child is given a small amount of the liquid three times a day from the left palm.

Family: *CONNARACEAE*
Botanical Name: *Cnestis ferruginea* DC.

Local Name: Wulukese (Ko)
Local Name:
Nyamawai (M)

Medicinal Use (1): Convulsion in children (Ko)
Preparation and Treatment (1): The leaves are boiled together with the leaves of *Microglossa pyrifolia*. About 50 ml is given to the child twice a day. Part of the filtrate can also be used to bathe the child once a day.
Medicinal Use (2): Gum disease (P)
Preparation and Treatment (2): The juice obtained from the fruits is applied dropwise to the affected gum to ease the pain.

Family: *CONVOLVULACEAE*
Botanical Name: *Ipomoea involucrate*
P. Beauv.

Local Name:
nDondokoei (M) or
Tofo-Mese (Ko)

Medicinal Use: Hemostasis of the nose (P) (Ko)
Preparation and Treatment: The juice obtained from the leaves is applied dropwise into the nostrils.

Family: *COSTACEAE/ZINGIFERACEAE*
Botanical Name: *Costus afer* Ker-Gawl

Local Name: Howae / Howoi (M)
Local Name: Tofa (Ko)

Use (1): Bleaching agent (Ke)
Preparation and Treatment (1): The external layer of the stem is scraped off and pounded, together with the internal soft tissue. Stained

material is washed, and the fluid is squeezed out of the stem tissue onto the stained area. After forty-five to sixty minutes, the material is washed with a soap.

Medicinal Use (2): Gonorrhea (Ke) (P)

Preparation and Treatment (2): The soft internal tissue of *Costus afer* stem and the bark of *Craterispernum laurinum* with the epidermal layer striped off are pounded together and then used to make a decoction with a few fruits of lime (*Citrus aurantifolia*) and drunk (Ke). A maceration of the pounded stem (without epidermis) with water and fruits of *Citrus aurantifolia* is drunk (P).

Medicinal Use (3): Yellow fever (Ko)

Preparation and Treatment (3): The stem is crushed in water and lime juice (*Citrus aurantifolia*) is added to the extracted juice. Approximately 50 ml is taken three times a day.

Family: *CRASSULACEAE*
Botanical Name: *Bryophyllum pinnatum* (Lam.) Oken

Local Name: Gba-Yamba (Ko)
Local Name: Kpolaa (M)
Local Name: Never die (K)

Medicinal Use (1): Convulsion (H)

Preparation and Treatment (1): A decoction of the leaves of *Bryophyllum pinnatum* and *Portulaca olefacea* is used to bathe babies.

Medicinal Use (2): Diseases of the ear/earache (G)(H)

Preparation and Treatment (2): The juice obtained from heated leaves, either undiluted or with a few drops of coconut oil added, is dropped into the ear.

Medicinal Use (3): Fever (H)

Preparation and Treatment (3): A decoction of the leaves together with those of a wild variety of *Amaranthus hybridus* is used to bathe the patient.

Medicinal Use (4): Healing of the navel wound in newborn babies (H) (Ko) (P)

Preparation and Treatment (4): The *Bryophyllum* leaves are heated; the juice is squeezed out and then mixed with salt and applied to the navel

(H). The juice obtained from the heated leaves is applied to the navel, and talcum powder added to heal the navel of newborns (Ko) (P).

Family: *CUCURBITACEAE*
Botanical Name: *Momordica charantia* Linn.   **Local Name:** Agerie (K)
Medicinal Use (1): Malaria (G) (H)
Preparation and Treatment (1): An infusion (G) or a decoction of the leaves together with those of *Morinda geminata* is drunk (150 ml, twice a day). (H)
Medicinal Use (2): Constipation in babies (H)
Preparation and Treatment (2): A decoction of the leaves is given to babies (one teaspoonful a day).

Family: *CYPERACEAE*
Botanical Name: *Scleria boivinii* Steud.   **Local Name:** Baboo
/*S.barteri* Boeck   razor (K)
Medicinal Use: To cure allay, a skin disease (H)
Preparation and Treatment: The leaf is made into a poultice with palm oil and applied to the affected areas.

Family: *DICHAPETALACEAE*
Botanical Name: *Dichapetalum toxicarium*   **Local Name:** Magbevei
(G. Don) Baill.   (M)
     Local Name: Brokobak (K)
Medicinal Use: Rat poison (Ke)
Preparation and Treatment: The mature fruits are crushed and applied to any food eaten by the rats, preferably rice and fish.

Family: *DILLENIACEAE*
Botanical Name: *Tetracera alnifolia* Willd.   **Local Name:** Katatei (M)
Medicinal Use (1): Dysentery (P)
Preparation and Treatment (1): Leaves are pounded with water and lime is added to the filtrate and drunk.

Medicinal Use (2): Thorn prick (P)
Preparation and Treatment (2): Exudates from the conducting tissues of the stem and a cataplasm of the leaves are applied topically to the affected area.

Family: *DILLENIACEAE*
Botanical Name: *Tetracera leicarpa* Stapf.   **Local Name:** Lepet tongue or Kondotail (K)

Medicinal Use (1): Gonorrhea and urinary tract infection (blockage)(H)
Preparation and Treatment (1): The leaves are pounded in a mortar; water and the juice from twenty-four lime fruits are added and drunk.
Medicinal Use (2): Cough (H)
Preparation and Treatment (2): The leaves together with chopped, dried, Cola nuts (*Cola nitida*) fruits are ground into a powder with *Aframomum* and honey added to form the drug. A small amount is administered from the palm.

Family: *DILLENIACEAE*
Botanical Name: *Tetracera potatoria* Afzel.   **Local Name:** Katatei (M)
Local Name: Kuin-Nenene (Ko)
Local Name: Lepet tongue (K)

Medicinal Use (1): Dysentery (Ke)
Preparation and Treatment (1): Leaves are macerated with a small amount of cold water. The concentrated green slurry is cleared of leaf tissue, and common salt is added. The liquid is then drunk.
Medicinal Use (2): Hemorrhoids (G)
Preparation and Treatment (2): An infusion of the leaves is drunk.
Medicinal Use (3): Tuberculosis (Ko)
Preparation and Treatment (3): Young leaves are pounded with beniseeds (*Sesamum indicum*), made into balls, and dried under the sun. A piece of the ball is eaten two to three times a day.

Family: *DIOSCOREACEAE*

Botanical Name: *Dioscorea bulbifera* Linn.   **Local Name:** Am-bonk
Local Name: Bush yam (K)

Medicinal Use (1): Aids childbirth (H)
Preparation and Treatment (1): Just before labor pains, the pregnant woman is bathed with a decoction of the plant and that of *Tetracera leicarpa.*
Medicinal Use (2): Fertility (H)
Preparation and Treatment (2): A decoction of the stem of the plant is drunk.
Medicinal Use (3): Headache (Ma)
Preparation and Treatment (3): Fresh panicles are rubbed over the forehead or dried leaves pounded with water and the macerate applied to the forehead.

Family: *EBENACEAE*
Botanical Name: *Diospyros heudelotii* Hiern.   **Local Name:** Tonyamba (Ko)
Local Name: nDoku-wui (M)

Medicinal Use (1): Stops vomiting and hiccups (Ko)
Preparation and Treatment (1): The leaves are boiled for about an hour. About 150 ml for adults and approximately 50 ml for children is taken as soon as hiccups or vomiting starts.
Medicinal Use (2) Skin diseases (P)
Preparation and Treatment (2) Young leaves are wrapped with larger ones and heated. The juice obtained by squeezing the former ones is applied to the affected areas. In addition the leaves can be ground into a poultice with white clay and applied to the affected areas.

Family: *EBENACEAE*
Botanical Name: *Diospyros thomasii* Hutch. & Dalz.   **Local Name:** N' Doku-Wulo—Leli (M)

Medicinal Use : Dysentery (Ke)
Preparation and Treatment: The bark of the stem is soaked in cold water for approximately two to three hours, and the brown liquid is then taken orally three times a day.

Family: *EUPHORBIACEAE*
Botanical Name: *Alchornea cordifolia*
(Schum. & Thonn.) Mull.-Arg
(Figure13)

Local Name:
An-thoma (T)
Local Name: Dui (Ko)
Local Name: Chrismes
stick (K)
Local Name: Krismes
Leaf (K)
Local Name: N' Jekoei
(M)

Medicinal Use (1): Abnormal bleeding in women (H)
Preparation and Treatment (1): A very concentrated decoction of the *Alchornea cordifolia* leaves is made and the patient sits in a bowl of the decoction twice a day.
Medicinal Use (2): Constipation in suckling mother (H)
Preparation and Treatment (2): The patient drinks about 200 ml of a decoction of the leaves once a day.
Medicinal Use (3): Cough (Ko)
Preparation and Treatment (3): Young leaves are boiled and the juice from lime fruits (*Citrus aurantifolia*) added, and approximately 150 ml of this decoction is drunk, twice a day.
Medicinal Use (4): Development of premature children (Ko)
Preparation and Treatment (4): The leaves are pounded in a mortar and slightly heated to lukewarm temperature. This is wrapped in clothing and put underneath the child for an hour or two a day.
Medicinal Use (5): Heart disease (H)
Preparation and Treatment (5): The epidermis of the bark of the *Alchornea cordifolia* plant is removed and the inner layer is chewed regularly.
Medicinal Use (6): Sore throat (Ko)
Preparation and Treatment (6): Three to four young leaves are chewed and the juice swallowed three times a day.
Medicinal Use (7): Stops bleeding of wounds (Ke)

Preparation and Treatment (7): Fluid from macerated leaves is applied to the wound.

Medicinal Use (8): Swelling of the foot (Ma)

Preparation and Treatment (8): The bark is pounded into a powder and moistened with cold water to form a cataplasm and applied to the limb. This causes water to exude from the limb.

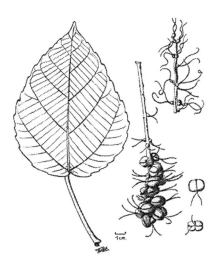

Figure 13—*Alchornea cordifolia*

Family: *EUPHORBIACEAE*
Botanical Name: *Erythrococca anomala* (Juss.) Prain   **Local Name:** Bush Lem (K)

Medicinal Use (1): Malaria (H)
Preparation and Treatment (1): A decoction of the roots is drunk (approximately 200 ml, twice a day).
Medicinal Use (2): Aids teething in babies (H)
Preparation and Treatment (2): A decoction of the root is drunk (one teaspoonful, once a day).

Family: *EUPHORBIACEAE*
Botanical Name: *Euphorbia deightonii* Croizat   **Local Name:** N' Gele-goni (M)
Local Name: Orro (K)

Medicinal Use: Inflammation (boils) (Ke)

Preparation and Treatment: The succulent leaves are macerated on a flat solid surface and then applied as a lotion.

Family: *EUPHORBIACEAE*
Botanical Name: *Jatropha curcas* Linn.   **Local Name:** Katakone
                                                    (Ko)
Medicinal Use (1): Yellow fever (Ko)
Preparation and Treatment (1): The leaves are boiled with the fruits of lime (*Citrus aurantifolia*). Approximately 150 ml of the drug is given to the patient in the morning and evening.
Medicinal Use (2): Sores (Ko)
Preparation and Treatment (2): Leaves are heated slightly to make them soft and crushed to make a soft paste. The soft paste is applied as a poultice on the site of the sore and bandaged.

Family: *EUPHORBIACEAE*
Botanical Name: *Macaranga* barteri   **Local Name:** Papu-Gbe
(Mull.—Arg)   (Ko)
Medicinal Use: Bilharzia (Ko)
Preparation and Treatment: Leaves are pounded in a mortar with water and filtered. Several lime fruits (*Citrus aurantifolia*) are divided into halves and added into the filtrate. Approximately 150 ml of the mixture is drunk three to four times a day.

Family: *EUPHORBIACEAE*
Botanical Name: *Macaranga heterophylla*   **Local Name:**
(Mull.—Arg)   Papu-Yawa (Ko)
Medicinal Use: Gonorrhea (Ko)
Preparation and Treatment: Leaves are pounded in a mortar with water and filtered. Several limes (*Citrus aurantifolia*) are divided into halves and the juice squeezed into the filtrate. Approximately 150 ml of the mixture is drunk three to four times a day.

Family: *EUPHORBIACEAE*
Botanical Name: *Manihot esculenta*   **Local Name:** Cassada
Crantz   leaf (K)
Medicinal Use (1): Eye problems (H)

Preparation and Treatment (1): The leaves are crushed between the palms and the juice applied into the eye, once a day.

Medicinal Use (2): Sores (H)

Preparation and Treatment (2): The tubers are pounded and made into a poultice using nut oil (from coconut) and then applied to the sore several times a day.

Family: *EUPHORBIACEAE*

Botanical Name: *Manniophyton fulvum* (Mull.—Arg)   **Local Name:** nJolei (M) or Yoo (Ko)

Medicinal Use (1): Dysentery (P) (Ko)

Preparation and Treatment (1): Young leaves together with beniseed (*Sesamum indicum*) are pounded, made into balls, dried in the sun and eaten, two to three times a day.

Medicinal Use (2): Cough (P) (Ko)

Preparation and Treatment (2): Stems without the epidermis are chewed and the juice swallowed several times a day.

Family: *EUPHORBIACEAE*

Botanical Name: *Mareya micrantha* (Benth.) Mull.Arg   **Local Name:** Nomba-Wan (K), Number One (K)
Local Name: Nonangona Nomba-wan (Ko)
Local Name: Nwanwa (M), nwanwai (M)

Medicinal Use (1): Laxative (P) (Ko)

Preparation and Treatment (1): A decoction of the young leaves is drunk.

Medicinal Use (2): Malaria (G)

Preparation and Treatment (2): An infusion of the dried leaves with those of *Morinda geminata*, *Cassia siberiana*, and *Momordica charantia* is drunk.

Medicinal Use (3): Measles (P) (Ko)

Preparation and Treatment (3): Heated leaves are made into a poultice with white clay and water, and evenly applied on the child. In addition, a decoction of the older leaves is used to bathe the child.

Medicinal Use (4): Skin disease, scabies (Ke)
Preparation and Treatment (4): A decoction of the leaves is used to bathe, and young leaves are macerated with a minimum amount of cold water and applied as a lotion.
Medicinal Use (5): Scabies (P) ("Craw Craw" in Krio)
Preparation and Treatment (5): The juice obtained from heated leaves is made into a poultice with white clay and applied to the affected region.
Medicinal Use (6): Stomach pains (P)
Preparation and Treatment (6): A decoction of the young leaves is drunk.

Family: *EUPHORBIACEAE*
Botanical Name: *Microdesmis puberula* Hook.f.ex Planch

Local Name: Antony leaf (K)
Local Name: Nikii (M)
Local Nmae: Ningine (Ko)

Medicinal Use (1): Hiccups (H)
Preparation and Treatment (1): The leaves are macerated and nut oil added, and two tablespoon full of the liquid is drunk.
Medicinal Use (2): Induce labor (H)
Preparation and Treatment (2): The leaves are macerated in a calabash, and the patient, while standing at a doorstep, drinks the liquid.
Medicinal Use (3): Nausea (H)
Preparation and Treatment (3): A decoction of the leaves is drunk and used to wash the face.
Medicinal Use (4): Scabies (P)
Preparation and Treatment (4): Leaves together with those of *Desmodium adscendens* are pounded and made into a poultice with white clay and applied to the affected area.
Medicinal Use (5): Itching skin (Ko)
Preparation and Treatment (5) The leaves together with those of *Desmodium adscendens* are pounded and white clay and water added to form a poultice, which is applied to the affected areas for 24 hours before bathing.

Family: *EUPHORBIACEAE*

Botanical Name: *Oldfieldia africana* Benth. & Hook.f. **Local Name**: Kpaola (M)

Medicinal Use: Insecticide (Ke)

Preparation and Treatment: The fumes obtained from maceration of fresh leaves are used to repel insects e.g. honeybee.

Family: *EUPHORBIACEAE*

Botanical Name: *Phyllanthus discoideus* (Baill.) Mull.-Arg **Local Name:** Tisoe (Ko)

Medicinal Use (1): Gum and toothache (Ko)

Preparation and Treatment (1): Leaves are boiled and the decoction obtained is the drug. Two to three mouthfuls are taken several times a day while the decoction is still lukewarm.

Medicinal Use (2): Diarrhea (Ko)

Preparation and Treatment (2): Leaves are boiled, and the decoction is the drug. Approximately 150 ml is taken three times a day.

Medicinal Use (3): Hemorrhoids (piles) (Ko)

Preparation and Treatment (3): The bark of the tree is boiled and the decoction used to make a sauce, or the decoction (~150 ml) can be drunk two or three times a day.

Family: *EUPHORBIACEAE*

Botanical Name: *Phyllanthus muellerianus* (O Ktze) Exell **Local Name:** Bush Beri or **Local Name:** Polly Mot Chuk Chuk (K) **Local Name:** Kundunomoe (Ko)

Medicinal Use (1): Dysentery (G)

Preparation and Treatment (1): The juice obtained from the leaves is drunk early in the morning before meals.

Medicinal Use (2): Gonorrhea (Ko)

Preparation and Treatment (2): The decoction obtained by boiling the root is the drug. About 150 ml is drunk three times a day.

Medicinal Use (3): Malaria (H)

Preparation and Treatment (3): A decoction of the leaves together with the leaves of *Chlorophora regia* and *Nauclea latifolia* is drunk.

Family: *EUPHORBIACEAE*
Botanical Name: *Ricinus communis* Linn.   **Local Name:** Caster oil
(Figure 14)   (K)
                                                   Local Name: An-fental
                                                   (T)

Medicinal Use (1): Headache (H)
Preparation and Treatment (1): The leaves are tied to the head of the patient once a day, which causes the patient to perspire.
Medicinal Use (2): Headache (Ma)
Preparation and Treatment (2): Dried leaves are heated and applied to the forehead.
Medicinal Use (3): Swelling of the leg (Ma)
Preparation and Treatment (3): Leaves are placed in hot water and the warm leaves wrapped round the leg. This process is repeated for fifteen to thirty minutes, after which the leg is dried and the patient put to bed, where leg will perspire profusely.

Figure 14—*Ricinus communis*

Family: *FABACEAE*
Botanical Name: *Abrus precatorius* Linn.   **Local Name:** Bodyay/
(Figure 15)   sweet leaf (K)
Medicinal Use: Cough (G)
Preparation and Treatment: An infusion of the leaves is drunk.

Medicinal Use: Cough (H)
Preparation and Treatment: A decoction of the leaves is made to which shea butter (*orie*) is added and drunk.

Figure 15—*Abrus precatorius*

Family: *FABACEAE*
Botanical Name: *Acacia pennata* Hutch. & Dalz.　　　　**Local Name:** Tarhan (L)
Medicinal Use: Toothache (G)
Preparation and Treatment: Spines and cork cells from the bark are removed and the green exposed layer mixed with palm oil and salt. This mixture is then inserted into the cavity for extraction of the tooth into pieces, care being taken to prevent contact with healthy teeth.

Family: *FABACEAE)*
Botanical Name: *Afzelia sp. (not africana).*　　　**Local Name:** Mapoor (T)
Medicinal Use: Headache (Ma)
Preparation and Treatment: Leaves are dried in the sun and pounded into a powder, then made into a cold cataplasm with water and applied to the forehead.

Family: *FABACEAE*
Botanical Name: *Albizia adianthifolia*　　　**Local Name:** Kpangba
(Schum.) Wright　　　(Ko)
Medicinal Use: Convulsion and fever (Ko)
Preparation and Treatment: The leaves are boiled, and the patient's head is covered over the vapor and the fumes from the hot drug are inhaled. In addition, the decoction is used for bathing, once a day.

Family: *FABACEAE*
Botanical Name: *Albizia zygia* (DC)　　　**Local Name:** Foo (Ko)
Medicinal Use: Dysentery (Ko)
Preparation and Treatment: The young leaves are boiled and pounded to make a soft paste with common salt (sodium chloride) added to give a taste. This drug is eaten with food e.g. rice by the patient.

Family: *FABACEAE*
Botanical Name: *Amphimas pterocarpoides*　　　**Local Name:**
Harms　　　Kuiyombo (Ko)
Medicinal Use (1): Stops bleeding and facilitates healing (Ko)
Preparation and Treatment (1): The bark of the tree is tapped and drops of the exuded latex applied to the site and firmly tied with a bandage.
Medicinal Use (2): Dysentery (Ko)
Preparation and Treatment (2): A decoction of the bark is made and ~150 ml drunk three times a day.

Family: *FABACEAE*
Botanical Name: *Anthonotha macrophylla* P.　　　**Local Name:** mBombi
Beauv.　　　(M)
Medicinal Use (1): Facilitate closure of the frontal suture in babies (Ke)
Preparation and Treatment (1): The young leaves are macerated and palm oil added and then applied to the area.
Medicinal Use (2): Yellow fever (G)
Preparation and Treatment (2): The juice obtained by macerating the leaves in water is drunk.

Family: *FABACEAE)*

Botanical Name: *Cajanus cajan* (Linn.) Millsp.   **Local Name:** Kongo Binch (K)

Local Name: So-Mese (Ko)

Medicinal Use (1): Cough (G)

Preparation and Treatment (1): An infusion of the leaves is drunk.

Medicinal Use (2): Measles (Ko)

Preparation and Treatment (2): A decoction of the leaves is drunk (~100 ml three times a day). Also the leaves are macerated with white clay and some water to make a soft paste, which is rubbed evenly on the patient's skin.

Medicinal Use (3): Measles (P)

Preparation and Treatment (3): A poultice obtained from pounded leaves is applied topically over the whole body.

Medicinal Use (4): Measles (G)

Preparation and Treatment (4): An infusion of the leaves is used to bathe and is also drunk.

Medicinal Use (5): Toothache (H)

Preparation and Treatment (5): The leaves are pounded in a mortar. With the patient standing at a door entrance, holding seven pebbles, the pounded leaves are placed in a pot of boiling water. The steam from the pot is allowed to heat the affected tooth and the lukewarm decoction is used to gargle, three times a day.

Family: *FABACEAE*

Botanical Name: *Cassia alata* Linn.   **Local Name:** nJepaa (M)

Local Name: N'Japai (M)

Local Name: Craw craw plant (K)

Medicinal Use (1): Laxative (P) (Ke)

Preparation and Treatment (1): A decoction of the dried leaves is drunk (P). A decoction of the leaves is taken orally (Ke).

Medicinal Use (2): Parasitic skin disease, Tinea—"Facincloth" (K)(Ke)

Preparation and Treatment (2): The leaves of *Cassia alata* are macerated into a finely homogenized paste and applied as a lotion.

Family: *FABACEAE)*
Botanical Name: *Cassia occidentalis* Linn.   **Local Name:** Stinkinwid, stinking weed (K)

Medicinal Use (1): Convulsion (H)
Preparation and Treatment (1): The *Cassica occidentalis* leaves are made into a poultice with palm oil and then applied over the patients body.
Medicinal Use (2): Cough (G)
Preparation and Treatment (2): An infusion of the leaves is drunk.
Medicinal Use (3): Diabetes (G) (H)
Preparation and Treatment (3): An infusion of the leaves is drunk (G). The seeds are dried, roasted, and pounded, and a decoction of this is drunk as a beverage (H)
Medicinal Use (4): Fever (G)
Preparation and Treatment (4): Pounded leaves are applied over the body.
Medicinal Use (5): Hypertension (G)
Preparation and Treatment (5): An infusion of the leaves is drunk.
Medicinal Use (6): Abnormal Heart Beat/Palpitations (H)
Preparation and Treatment (6): The *Cassia occidentalis* seeds are dried, roasted, and pounded, and a decoction of this is drunk as a beverage.

Family: *FABACEAE*
Botanical Name: *Cassia podocarpa (Guill. & Perr.)*   **Local Name:** Teteteteh (M)
Medicinal Use: Head lice (G)
Preparation and Treatment: A powder obtained from burnt leaves is mixed with local alum *(lubi)* and the mixture applied to the head.

Family: *FABACEAE)*
Botanical Name: *Cassia siamea* Lam.   **Local Name:** Seku Turay (K) (Ko) or Sheku Turay (K)
Medicinal Use: Malaria (P) (Ko) (H)
Preparation and Treatment: A maceration or a decoction of the root is drunk (P).

Also, the roots may be cut into small pieces and put in a container or bottle, and water added to the brim; or the roots may be boiled and the decoction drunk (Ko); or a decoction of the leaves is drunk (H).

Family: *FABACEAE)*
Botanical Name: *Cassia siberiana* DC.　　**Local Name:** Ganya (M), Gbangba (M)
Local Name: Gbangba tik (K)
Local Name: Kukai (Ko)

Medicinal Use (1): Malaria (G)
Preparation and Treatment (1): A decoction of the roots with those of *Landolphia dulcis* and *Uapaca guineensis* is drunk.
Medicinal Use (2): Malaria and purgative (Ke)
Preparation and Treatment (2): A decoction of the roots is drunk.
Medicinal Use (3): Constipation (Ko)
Preparation and Treatment (3): The leaves are boiled and the decoction is drunk.
Medicinal Use (4): Weakness and pains (Ko)
Preparation and Treatment (4): The leaves are boiled and the decoction drunk two times a day.

Family: *FABACEAE*
Botanical Name: *Desmodium adscendens* (Sw)　**Local Name:**
DC.　　nDogbo-nikii (M)
Local Name: N' Dogbo-Niki (M)

Medicinal Use (1): Asthma (P)
Preparation and Treatment (1): A maceration of the leaves together with those of *Corchorus olitorius* is drunk. In addition the patient is massaged with the crushed leaves.
Medicinal Use (2): Reduce hard labor pain (Ke)
Preparation and Treatment (2): Leaves are macerated with a small amount of water, and the liquid is drunk.
Medicinal Use (3): Scabies (P)

Preparation and Treatment (3): Leaves together with those of *Microdesmis puberula* and white clay are made into a poultice and applied to the affected region.

Medicinal Use (4): Snake bite (Ke)

Preparation and Treatment (4): Macerated leaves are squeezed and the fluid dropped onto the bitten area and the area is surrounded with a soft paste of the leaves (except the wound). This helps to get the snake tooth out.

Family: *FABACEAE*

Botanical Name: *Dialium guineense* Willd.   **Local Name:** Blak tombla (K)

Local Name: Manbui (M)

Medicinal Use (1): Cough (G)

Preparation and Treatment (1): An infusion of the leaves with lime is drunk.

Medicinal Use (2): Gonorrhea (P)

Preparation and Treatment (2): A decoction of the leaves and branches is drunk.

Family: *FABACEAE*

Botanical Name: *Dichrostachys glomerata* (Forssk.) Chiov.   **Local Name:** Dandae/ nDandei (M),

Local Name: Pintikili (K)

Medicinal Use: Snake repellent (Ke)

Preparation and Treatment: The fibrous bark is tied around the ankles to deter the bites of snakes.

Medicinal Use (1): Snake repellent (H)

Preparation and Treatment (1): The bark of the plant is dried and made into a powder, and the powder is scattered around the house.

Medicinal Use (2): Worms (Ke)

Preparation and Treatment (2): The epidermal layer is discarded from the bark, and a decoction of the inner fibrous layer is made and taken orally.

Medicinal Use (3): Hemorrhoids (piles) (Ke) (P)

Preparation and Treatment (3): The fresh leaves together with the leaves of *Mareya micrantha* are macerated with a small amount of water and then applied externally to the affected area (Ke). Young leaves together with those of *Mareya micrantha* are ground with water and made into a poultice, which is applied to the anus (P).

Medicinal Use (4): Body pain (H)

Preparation and Treatment (4): A decoction of the leaves together with those of *Chlorofora regis* and *Nauclea latifolia* is drunk. In addition, the decoction can be added to the water when bathing.

Family: *FABACEAE*
Botanical Name: *Erythrophleum guineense* G. Don   **Local Name:** Sasswood (K)
Medicinal Use (1): Rheumatism (F)
Preparation and Treatment (1): Charcoal made from the bark is mixed with white clay and applied to the affected areas.
Medicinal Use (2): Rat poison (K)
Preparation and Treatment (2): The pounded bark is used.

Family: *FABACEAE*
Botanical Name: *Leptoderris trifoliata Hepper*   **Local Name:** Kawa-Kawa Wumboi (Ko)
Medicinal Use: Heal wounds (Ko)
Preparation and Treatment: The bark of the plant is scraped and the exuding sap collected and applied to the fresh wound, and replaced twice daily.

Family: *FABACEAE*
Botanical Name: *Mezoneurum benthamianum* Baill.   **Local Name:** Boge-Saloe (Ko)
Medicinal Use (1): To avert bedwetting (Ko)
Preparation and Treatment (1): Five or six young leaves are chewed and the juice swallowed three or four times a day.
Medicinal Use (2): Aid to increase potency (Ko)
Preparation and Treatment (2): Young leaves are pounded with local white rice, and a spoonful of this mixture is eaten every morning.

Family: *FABACEAE*
Botanical Name: *Mimosa pudica Linn.*
(Figure 16)

Local Name: Buyaki
(Ko)
Local Name: gBagbemi
Local Name: Gba—
gbema (M)
Local Name: Tie You
Lappa, You Man Dae
Cam (K)

Medicinal Use (1): Headache (Ma)
Preparation and Treatment (1): The leaves and small branches are bruised and applied as a poultice to the forehead.
Medicinal Use (2): Inflammation (P) (Ko)
Preparation and Treatment (2): A poultice obtained by macerating young leaves with goat fat cures skin inflammation.
Medicinal Use (3): Impotency (H)
Preparation and Treatment (3): The dried leaves are immersed in water, and a red hot axe is placed in the water and then filtered. The filtrate is then consumed with porridge or *"agidi"* (a maize meal).
Medicinal Use (4): Protection against evil spirits (H)
Preparation and Treatment (4): A needle is inserted into a white kola nut (*Cola nitida*) and then placed at the bottom of a pot containing water; the leaves of *Mimosa pudica* together with the leaves of *Thaumatococcus daniellii* are placed in the water and boiled. The decoction is then drunk.
Medicinal Use (5): Parasitic skin disease, *Tinea versicolor*, "Facinclot"(K) (Ke)
Preparation and Treatment (5): Macerated leaves are applied as a lotion.
Medicinal Use (6): Pain
Preparation and Treatment (6): A decoction is used to bathe face and gargle gums when swollen and painful.

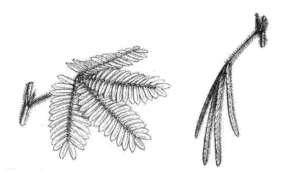

Figure 16—*Mimosa pudica*

Family: *FABACEAE*
Botanical Name: *Phaseolus lunatus* Linn.   **Local Name:** Binch (K)
Local Name: Towei (M)
Medicinal Use: Ringworm (Ke)
Preparation and Treatment: Extract from macerated leaves are applied to the infected areas after scraping the area with *Ficus exasperata* leaves.

Family: *FABACEAE*
Botanical Name: *Tamarindus indica* Linn.   **Local Name:** Sour tumbla (K)
Local Name: Tombi (K)
Medicinal Use (1): Constipation (F)
Preparation and Treatment (1): The sweet fruit is eaten as a snack or rolled into balls and boiled alone or with a cereal beverage, and drunk for constipation.
Medicinal Use (2): Wounds (F)
Preparation and Treatment (2): A decoction of the leaves is used to wash wounds; or the dried pounded leaves are used for sores.
Medicinal Use (3): Boils (F)
Preparation and Treatment (3): The young leaves are made into a poultice and applied to the boil.
Medicinal Use (4): Hiccups (H)

Preparation and Treatment (4): A decoction of the leaves together with the fruits of *Solanum melongena* is sipped.

Family: *FABACEAE*
Botanical Name: *Tephrosia vogelii* Hook.f.   **Local Name:** Tawui (M)
Medicinal Use: Fish poison (Ke)
Preparation and Treatment: Leaves made into a paste or powder is used to paralyse fishes, which are then easily caught.

Family: *FLACOURTIACEAE*
Botanical Name: *Caloncoba brevipes* (Stapf) Gilg   **Local Name:** Gene Wui (M)
Medicinal Use: Acute stomach pain (adults) (Ke)
Preparation and Treatment: A decoction of the roots of *Craterisperinum laurinum* together with the roots of *Ficus exasperata*, lime (*Citrus aurantifolia*) fruits, and the large leaves of *Setaria chevalieri* is drunk.

Family: *FLACOURTIACEAE*
Botanical Name: *Caloncoba echinata* (Oliv.) Gilg   **Local Name:** Goli (M)
Medicinal Use: Toothache (P)(F)
Preparation and Treatment: A maceration of the roots without the epidermis is applied dropwise through the nostril on the opposite side of the toothache.

Family: *GRAMINEAE*
Botanical Name: *Cymbopogon citratus*   **Local Name:** Lemon grass (K)
Medicinal Use (1): Fever (G)
Preparation and Treatment (1): A tea is made from the leaves and drunk.

Family: *GUTTIFERAE (CLUSIACEAE)*
Botanical Name: *Garcinia kola* Heckel   **Local Name:** Bitakola (K)
Medicinal Use (1): Cough (G)
Preparation and Treatment (1): A maceration of the sap is drunk.
Medicinal Use (2): Increase libido

Preparation and Treatment (2): Seeds are chewed to increase sexual urge.

Family: *GUTTIFERAE (CLUSIACEAE)*
Botanical Name: *Harungana madagascariensis*
Lam. ex Poir (Figure 17)

Local Name: Blod tri (K)
Local Name: Blood tri (K),
Local Name: Palmine tick (K)
Local Name: M' Bel ei (M)
Local Name: Sungbali (Ko)

Medicinal Use (1): Gonorrhea (Ko)
Preparation and Treatment (1): Three to four young leaves are chewed with salt (sodium chloride) three times a day.
Medicinal Use (2): Malaria (G)
Preparation and Treatment (2): An infusion of the dried leaves with those of *Musa sp.* is used to bathe.
Medicinal Use (3): Ringworm (Ke)
Preparation and Treatment (3): The orange-red resinous sap exuded from the inner bark layer of incised tree is applied, after scraping the infected area with *Ficus exasperata* leaves.
Medicinal Use (4): Stomach poison (H)
Preparation and Treatment (4): A decoction of *H. madagascariensis* with *Nauclea latifolia, Chlorophora regia,* and spice (*Xylopia*) is drunk.
Medicinal Use (5): Strengthens the body of pregnant women (H)
Preparation and Treatment (5): A decoction of the leaves together with *Newbouldia laevis* leaves and a white egg is used to bathe pregnant women at three months and at eight months of pregnancy.
Medicinal Use (6): Tonic (G)
Preparation and Treatment (6): An infusion of the fresh leaves is drunk.
Medicinal Use (7): Toothache (Ko)
Preparation and Treatment (7): Young leaves are boiled for about an hour. Patient inhales steam from the drug through the mouth two to three times a day, and a lukewarm decoction is put in the mouth, kept there for a minute or two, and spat out.

Family: *GUTTIFERAE (CLUSIACEAE)*
Botanical Name: *Mammea Americana* Linn. **Local Name:**
Dindomawo (Ko)
Local Name: Mammy
Apple (K)
Medicinal Use (1): Dysentery (Ko)
Preparation and Treatment (1): Three to four young leaves are chewed with kola nuts (*Cola nitida*) and swallowed, three times a day.
Medicinal Use (2): Scorpion bite (Ko)
Preparation and Treatment (2): Leaves are macerated and made into a soft paste, and applied to the site of the bite.

Family: *GUTTIFERAE (CLUSIACEAE*
Botanical Name: *Vismia guineensis* (Linn.) **Local Name:**
Choisy
mBeli-mBambei (M)
Local Name: Woman
Blod lif (K)
Medicinal Use (1): Antidote (nullifies effect of poisonous substances)
Preparation and Treatment (1): Fresh leaves are chewed or the leaf poultice is immersed in water and the filtrate drunk.
Medicinal Use (2): Blood tonic for children (P)
Preparation and Treatment (2): A decoction of the leaves together with those of *Canthium glabriflorum* is drunk.

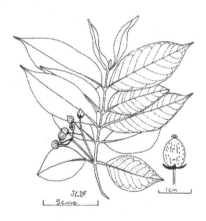

Figure 17—*Harungana madagascariensis*

Family: *LAMIACEAE (LABIACEAE)*

Botanical Name: *Hyptis suavolens* Poit.   **Local Name:** Bush Ti-bush (K)

Medicinal Use (1): Fever (F)

Preparation and Treatment: An infusion of the leaves is drunk.

Medicinal Use (2): Mosquito repellant(F)

Preparation and Treatment: The plant is placed in the room at night to repel mosquitoes

Family: *LAMIACEAE (LABIACEAE)*

Botanical Name: *Leonotis nepetifolia* (L.) R. Br.   **Local Name:** Nebul lif (K)

Medicinal Use: Yellow fever (G)

Preparation and Treatment: An infusion of the leaves is used to bathe.

Family: *LAMIACEAE (LABIACEAE)*

Botanical Name: *Ocimum americanum* Linn.   **Local Name:** Patmenji (K)

Local Name: Sasmenje (Pasmanije) (M)

Medicinal Use (1): Convulsion (P)

Preparation and Treatment (1): Crushed leaves are placed in the anus and the juice applied to the head, eyes, and mouth.

Medicinal Use (2): Lower abdominal pain (H)

Preparation and Treatment (2): A decoction of the leaves is drunk while hot.

Medicinal Use (3): Stops vomiting (H)

Preparation and Treatment (3): A decoction of the leaves is sipped once cooled.

Family: *LAMIACEAE (LABIACEAE)*

Botanical Name: *Ocimum basilicum* Linn.   **Local Name:** Patmange (K)

Medicinal Use (1): Convulsion (F)

Preparation and Treatment (1): The leaves are crushed and the juice applied dropwise to the head, eyes, arms, and mouth.

Medicinal Use (2): Menstrual pain (F)

Preparation and Treatment (2): An infusion of the leaves is drunk.

Medicinal Use (3): Piles (F)
Preparation and Treatment (2): An infusion of the leaves is drunk.
Medicinal Use (4): Fever (F)
Preparation and Treatment (4): An infusion of the leaves is drunk.

Family: *LAMIACEAE (LABIACEAE)*
Botanical Name: *Ocimum gratissimum* Linn.   **Local Name:** Tibush (K)
Local Name: Kumui (Ko)
Medicinal Use (1): Convulsion (Ko)
Preparation and Treatment (1): Leaves are crushed, and the juice that is obtained on squeezing is the drug. Two to three drops of the juice are dropped into the nostrils once a day.
Medicinal Use (2): Dysentery (G)
Preparation and Treatment (2): After crushing leaves, water and salt are added and the filtrate drunk.
Medicinal Use (3): Worms (Ko)
Preparation and Treatment (3): The decoction of the leaves is the drug. Approximately 150 ml is drunk as the first drink in the morning and the last drink in the evening.

Family: *LAMIACEAE (LABIACEAE)*
Botanical Name: *Ocimum viride* Willd.   **Local Name:** Teabush (K)
Local Name: Kumui (M)
Medicinal Use (1): Convulsion (P)
Preparation and Treatment (1): Crushed leaves are placed in the anus of babies; in addition, the juice from the crushed leaves is applied dropwise to the head, eyes, and mouth.
Medicinal Use (2): Laxative (P)
Preparation and Treatment (2): A decoction of the leaves is drunk.
Medicinal Use (3): Vaginal infection (H)
Preparation and Treatment (3): The macerated leaves are placed between the vaginal labia. Also, salt, nut oil, and water can be added to the macerated leaves and heated in a fireplace before applying to the site.
Medicinal Use (4): Stomach pains (H)

Preparation and Treatment (4): Pounded leaves are placed in lukewarm saline solution and filtered. The filtrate is the drug.

Family: *LAURACEAE*
Botanical Name: *Cinnamomum zeylanicum* Ness
Local Name: Cinnamon (K)
Medicinal Use: Headache (F)
Preparation and Treatment: A tea of the bark is drunk or the leaves are crushed in water and lime, and applied to the forehead.

Family: *LAURACEAE*
Botanical Name: *Persea americana* Mill.
Local Name: Pia or avocado pear (K)
Medicinal Use: Malaria (G) (H)
Preparation and Treatment: An infusion of the leaves with those of *Momordica charantia, Morinda geminata,* and *Carica papaya* is used to bathe (G). A decoction of the pounded leaves and roots together with the leaves of *Nauclea latifolia* is drunk (H).

Family: *LOGANIACEAE*
Botanical Name: *Anthocleista nobilis* G Don
Local Name: Pongui (M)
Local Name: Pongoi Henei (M)
Medicinal Use (1): Closure of frontal suture (P)
Preparation and Treatment (1): A poultice obtained by macerating leaves with water and soil from fireplace is applied to the forehead.
Medicinal Use (2): Laxative (P)
Preparation and Treatment (2): A decoction of the roots is drunk.
Medicinal Use (3): Purgative (Ke)
Preparation and Treatment (3): A decoction of the leaves is drunk.

Family: *LOGANIACEAE*
Botanical Name: *Anthocleista vogelii* Planch.
Local Name: An-kep-kep (T)
Local name: Pongo (L/M
Medicinal Use: Pelvic pain (G)

Preparation and Treatment: A decoction of the bark is drunk.

Family: *LOGANIACEAE*
Botanical Name: *Strychnos afzelia* Gilg   **Local Name:** Siminji (M)
Medicinal Use 1: Bad breath (Ma)
Preparation and Treatment 1: The leaves are chewed; this also makes the lips red.
Medicinal Use 2: Febrifuge for fever (Ma)
Preparation and Treatment 2: A decoction of the seeds together with leaves of *Ocimum sp.* is drunk.
Medicinal Use 3: Stomachache (Ma)
Preparation and Treatment 3: Ground seeds are added to a soup.
Medicinal Use 4: Fragrance and aphrodisiac
Preparation and Treatment 4: Women rub the juice from crushed leaves, pure or mixed with white clay, on their bodies as a fragrance and as an aphrodisiac.

Family: *LORANTHACEAE*
Botanical Name: *Tapinanthus bangwensis*   **Local Name:** Afomor (Engl. & K Krause)   (K)
Medicinal Use (1): Rheumatic pain or body pain (H)
Preparation and Treatment (1): The leaves of the plant together with the leaves of *Scleria boivinii* and the roots of *Elaeis guineensis* are boiled and the steam used to heat the painful area.
Medicinal Use (2): Stomach pains (H)
Preparation and Treatment (2): A decoction of the leaves together with the leaves of *Amaranthus hybridus var. cruentus* is drunk.

Family: *MALVACEAE*
Botanical Name: *Gossypium hirsutum* Linn.   **Local Name:** Wik (K), Wik leaf (K)
Medicinal Use (1): Causes a disease called "Wanka" on those who attempt to steal from a farm (H)
Preparation and Treatment (1): The bark of *Nauclea latifolia* is placed in a calabash. The cotton from *Gossypium hirsutum* is placed at the bottom of a calabash, which is then drilled; the calabash and its contents are then suspended in the middle of the farm.

Medicinal Use (2): Dysentery (G)
Preparation and Treatment (2): The juice obtained from pounded leaves or a decoction of the leaves is drunk.

Family: *MALVACEAE*
Botanical Name: *Hibiscus esculentus* Linn.　　**Local Name:** Bonde (Ko)
　　Local Name: Bondei (M)
Medicinal Use: Helps to close the frontal suture in new born babies (Ko), (P)
Preparation and Treatment: The leaves are crushed with a few drops of water until a soft paste is formed. The paste is applied as a liniment to the frontal suture of the head, once a day and for at least five hours (Ko) or once or twice a day (P)

Family: *MALVACEAE*
Botanical Name: *Hibiscus sabdariffa* Linn.　　**Local Name:** Sawa/ Sorrel (K)
Medicinal Use: Impotence (H)
Preparation and Treatment: A macerate of the leaves is used to massage the penis, after which a bath is taken with a local black soap. This treatment is administered twice a day.

Family: *MALVACEAE*
Botanical Name: *Hibiscus surattensis* Linn.　　**Local Name:** Bush sawa (K)
Medicinal Use: Cough (G)
Preparation and Treatment: An infusion of the leaves together with those of *Zingiber officinale* or *Ocimum gratissimum* is drunk.

Figure 18—*Hibiscus rosa-sinensis*

Family: *MALVACEAE*
Botanical Name: *Sida stipulata* Cav.   **Local Name:** Bush
Crain-Crain (K)
Local Name: Helui/
Helu (M)

Medicinal Use (1): Blue boils (Ke)
Preparation and Treatment (1): A soft paste is obtained from macerated leaves and then applied as a lotion to the affected areas.
Medicinal Use (2): Boils (H)
Preparation and Treatment (2): The leaves are mixed with palm oil and salt to form a poultice, which is applied twice a day. (The word "salt" should not be mentioned during the preparation)
Medicinal Use (3): Contraceptive (H)
Preparation and Treatment (3): The outer epidermis of the bark of *Sida stipulata* plant is peeled off and the clean inner bark twisted around a ring of black thread. The patient declares her wishes during the preparation of the drug. The contraceptive ring is then tied around the hips immediately after a menstrual period or after the birth of a child for as long as is desirable.
Medicinal Use (4): Inflammation (P)
Preparation and Treatment (4): A poultice obtained by crushing the leaves is applied to the affected region.

Family: *MELASTOMATACEAE*
Botanical Name: *Dissotis rotundifolia* (Sm.)   **Local Name:**
Triana   nGaku-wui (M)
Medicinal Use (1): Cough (P)
Preparation and Treatment (1): A concentrated decoction of the leaves
is drunk.
Medicinal Use (2): Frequent urination (P)
Preparation and Treatment (2): The filtrate obtained from macerated
leaves is drunk.
Medicinal Use (3): Gonorrhea (P)
Preparation and Treatment (3): A decoction of the leaves is drunk.
Medicinal Use (4): Eye problems (P)
Preparation and Treatment (4): The filtrate obtained from crushed
leaves is used as an eye drop.

Family: *MELIACEAE*
Botanical Name: *Azadirachta indica* A Juss.   **Local Name:** Kwinin
                                                     lif (K)
Medicinal Use: Malaria (G)
Preparation and Treatment: A decoction of the leaves from *Azadirachta
indica* (neem tree) is drunk.

Family: *MELIACEAE*
Botanical Name: *Carapa procera DC.* (figure   **Local Name:** Kundi
19)   (K) or Krab wood (K)
Medicinal Use (1): Malaria (G)
Preparation and Treatment (1): An infusion of the leaves is drunk.
Medicinal Use (2): Laxative and upset stomach (H)
Preparation and Treatment (2): A decoction of the bark from the plant
is drunk.
Medicinal Use (3): Rheumatism (F)
Preparation and Treatment (3): The oil from the seeds is massaged on
the affected region.

Figure 19—*Carapa procera*

Family: *MELIACEAE*
Botanical Name: *Trichilia heudelotii* Planch.   **Local Name:** Jawei (K)
Medicinal Use: Cough (P)
Preparation and Treatment: A maceration of the pounded bark is drunk.

Family: *MELIANTHACEAE*
Botanical Name: *Bersama abyssinica* Fres.   **Local Name**: Jondobei (M)
Medicinal Use: Intestinal nematode worms (Ke)
Preparation and Treatment: The pounded bark of the stem is mixed with clay and rolled into balls and strips. The dried balls are ingested as the drug

Family: *MENISPERMACEAE*
Botanical Name: *Chasmanthera dependens*   **Local Name:** Flawa (K)
Hochst.
Medicinal Use: Boils (G)
Preparation and Treatment: The juice obtained by crushing the leaves is applied to the boil.

Family: *MENISPERMACEAE*

Botanical Name: *Dioscoreophyllum Spp.*

Local Name: Awayato (K)

Medicinal Use: Body pain

Preparation and Treatment: The leaves are bandaged over the affected area for an hour

Family: *MENISPERMACEAE*

Botanical Name: *Stephania dinklagei* (Engl.) Diels

Local Name: Ra-nes (T)

Medicinal Use: Fish poison (Ma)

Preparation and Treatment: The stem is pounded and placed on the water.

Family: *MENISPERMACEAE*

Botanical Name: *Triclisia patens* Oliv.

Local Name: Foklo-bei (M)

Medicinal Use: Stomach pain (Ke)

Preparation and Treatment: A decoction of the roots of *Triclisia patens* and the fruits of *Citrus aurantifolia* is drunk.

Family: *MORACEAE*

Botanical Name: *Chlorophora regia* A Chev.

Local Name: Iroko (K)

Medicinal Use: Strengthens feeble children (H)

Preparation and Treatment: A decoction of *Chlorophora regia* and *Nauclea latifolia* leaves is used to bathe the child, once a day.

Family: *MORACEAE*

Botanical Name: *Ficus capensis* Thunb.

Local Name: Waytewata (K)
Local Name: Wisi (Ko)
Local Name: nDahie (M)

Medicinal Use (1): Anemia (Ko)

Preparation and Treatment (1): The bark of the tree is removed, and the inner tissue boiled until a bright red liquid is obtained. This decoction is used to prepare a soup, with fresh meat or fish. The soup is eaten daily at meal times.

Medicinal Use (2): Closure of frontal suture (P)

Preparation and Treatment (2): Leaves made into a poultice with white clay and water facilitates closure of the frontal suture in babies.

Medicinal Use (3): Splitting of the edges of the feet

Preparation and Treatment (3): The juice from the fruit or the exudate from the bark is applied to the affected region.

Family: *MORACEAE*
Botanical Name: *Ficus exasperate* Vahl
(Figure 20)

Local Name: Kamaamei (M),
Local Name: Kama-Kama (Ko)
Local Name: Krach-lif (K)

Medicinal Use (1): Abortion (P) (Ko), abortion of young pregnancy (Ke)

Preparation and Treatment (1): The leaves are crushed in water in a calabash and three glowing charcoals added; the filtrate is the drug. Approximately 150 ml of filtrate is taken three times a day (P/Ko). Leaves are macerated with a little water and a glowing charcoal added to act as an adsorbent. The liquid is then drunk (Ke).

Medicinal Use (2): Convulsion (H)

Preparation and Treatment (2): The bark from the plant is used to make a rope, and the rope washed with a suspension made from *Amaranthus hybridus var. cruentus* and *Portulaca oleracea*. The rope is then tied around the patient's neck for six days.

Medicinal Use (3): Eruptive skin disease (called *Nyagofoi* in Mende) (P)

Preparation and Treatment (3): The leaves together with those of *Dissotis rotundifolia* and *Selaginella myosurus* are made into a poultice with white clay and applied to the affect area to cure eruptive skin disease.

Medicinal Use (4): Itching (Ko) and vaginal rash (P)(Ko)

Preparation and Treatment (2): The powder obtained from macerated dried leaves is applied to the affected area.

Medicinal Use (4): Ringworm (G) (Ke)

Preparation and Treatment: The juice from the young leaves is squeezed onto the affected region (G). The infected area is scraped with the leaf (Ke).

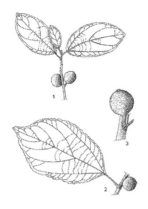

Figure 20—*Ficus exasperata*

Family: *MORACEAE*
Botanical Name: *Ficus umbellata* Vahl.　　　**Local Name:** Brod Lif (K)
Medicinal Use (1): Headache (H)
Preparation and Treatment (1): A decoction (~200 ml) of *Ficus umbellata* and *Newbouldia laevis* leaves is drunk three times a day.
Medicinal Use (2): Severe constipation and fever (H)
Preparation and Treatment (2): A decoction (~100 ml) of *Ficus umbellata* leaves to which epsom salt is added is drunk two times a day.
Medicinal Use (3): Malaria (H)
Preparation and Treatment (3): A decoction (~200 ml once a day) of *Ficus umbellata* leaves together with pounded bark of *Craterispermum laurinum*, leaves of *Nauclea latifolia*, and *Morinada geminata*, to which ~20 ml of lime juice is added, is drunk.

Family: *MORACEAE*
Botanical Name: *Musanga cecropioides* **R. Br. & Tedlie**　　　**Local Name:** Wunson (Ko)
Medicinal Use: Toothache (Ko)

Preparation and Treatment: The bark is removed and boiled. The lukewarm soft tissue and the filtrate is the drug. The lukewarm soft tissue of the bark is pressed on the spot of the ache and the decoction is used as a gargle several times a day.

Family: *MORACEAE*
Botanical Name: *Myrianthus serratus* (Trecul)   **Local Name:** Fofoi (M)
Benth. & Hook
Medicinal Use: Tonic (P)
Preparation and Treatment: A decoction of the bark is drunk by children to make them healthy and strong.

Family: *MUSACEAE*
Botanical Name: *Musa paradisiaca* Linn.   **Local Name:** E—Planti (T)
Local Name: Plantin (K)
Medicinal Use: Headache (Ma)
Preparation and Treatment: The leaves are heated and applied to the forehead.

Family: *MUSACEAE*
Botanical Name: *Musa sapientum* Linn.   **Local Name:** Banana (K)
Medicinal Use (1): Hiccups (H)
Preparation and Treatment (1): A decoction of the stalk is made; **honey added**, and administered orally from the right palm.
Medicinal Use (2): Stomach poison (H)
Preparation and Treatment (2): The fruit is mixed with nut oil and drunk to help vomit the poison.

Family: *MYRTACEAE*
Botanical Name: *Psidium guajava* Linn.   **Local Name:** Goyavei (M)
Local Name: Gugva (Ko)
Local Name: Gueva (K),
Local Name: Gweva (K),
Medicinal Use: Dysentery (G) (Ko), diarrhea and dysentery (Ke)

Preparation and Treatment: A decoction of the pounded leaves together with fruits of *Xylopia aethiopica* (spice) and *Citrus aurantifolia* is drunk (G). Three to four young leaves are chewed with common salt and swallowed once or twice a day, or the decoction obtained from the boiled leaves is the drug. Approximately 150 ml of this drug is drunk three times a day (Ko). A decoction of the leaves is taken orally (Ke).

Family: *OCHNACEAE*
Botanical Name: *Lophira lanceolata* van Triegh. **Local Name:** E-ninka
ex Keay   (T)
Medicinal Use: Gonorrhea (Ma)
Preparation and Treatment: An infusion of the leaves is injected through a reed by blowing the drug into the urethra.

Family: *OCHNACEAE*
Botanical Name: *Ouratea morsonii* Hutch. &   **Local Name:** Jaygay lif
Dalziel   (K)
Medicinal Use: Pelvic pain (G)
Preparation and Treatment: Leaves are crushed to form a poultice, which is bandaged around the pelvic region.

Family: *ORCHIDACEAE*
Botanical Name: *Calyptrochilum emarginatum* **Local Name:**
(Sw.) Schltr.   Gbogbo-Njamba (Ko)
Medicinal Use: To smooth protrusion in vagina due to inflammation (Ko)
Preparation and Treatment: The leaves are pounded and boiled; the decoction is then used to douche twice a day.

Family: *PALMAE (ARECACEAE)*
Botanical Name: *Ancistrophyllum secundiflorum* **Local Name:** Kangane
(P Beauv.) Wendl   (Ko)
Medicinal Use: For easy birth; believed to ease pain during the movement of the fetus (Ko)
Preparation and Treatment: Root and stem are cut into small pieces, and a few young leaves from the cotton tree (*Ceiba pentandra*) are added and boiled and the decoction drunk. The pregnant woman begins to take the drug as soon as labor pains are experienced. Approximately 150 ml is drunk twice a day.

Family: *PALMAE (ARECACEAE)*
Botanical Name: *Elaeis guineensis* Jacq.

Local Name: Banga-tik (K)
Local Name: Pam tri (K)
Local Name: Tokpoi (M)

Medicinal Use (1): Blue boils (P)
Preparation and Treatment (1): A poultice obtained by mixing burnt toad with boiling palm oil is applied to the boil.
Medicinal Use (2): Hiccups (H)
Preparation and Treatment (2): The bark of the plant is cut at the rising or setting of the sun, and a decoction of this is sipped until the hiccups stop.
Medicinal Use (3): Inflammation (P)
Preparation and Treatment (3): A poultice obtained by mixing burnt toad with boiling palm oil is applied to the boil.
Medicinal Use (4): Toothache (G) (Ke)
Preparation and Treatment (4): A decoction of the bark is used as a gargle (G). A decoction made from the bark of the stem is drunk at intervals while lukewarm (Ke).

Family: *PALMAE (ARECACEAE)*
Botanical Name: *Raphia vinifera* P. Beauv.   **Local Name:** Kejui (M)
Medicinal Use: Fish poison (Ke)
Preparation and Treatment: The stem or fruits are pounded to produce fibrous particles, which are thrown into a pool or at the back of a temporary dam. In high concentrations, the fishes float on the water surface, where they are easily collected.

Family: *PASSIFLORACEAE*
Botanical Name: *Adenia lobata* (Jacq.) Engl.   **Local Name:** Jumukei (M)
Medicinal Use: Fish poison (Ke)
Preparation and Treatment: The stem is pounded to produce fibrous particles, which are thrown into the water. In high concentrations, the fishes float on the water surface, where they are easily collected.

Family: *PASSIFLORACEAE*
Botanical Name: *Smeathmannia pubescens*   **Local Name:** nDovotei
Sol. ex R.Br.   (M)
Medicinal Use: Laxative (P)
Preparation and Treatment: A maceration of the roots is drunk (or for increased effectiveness, the roots of *Sarcocephalus esculentus* and *Morinda geminata* are included).

Family: *PEDALIACEAE*
Botanical Name: *Sesamum indicum* (L.)   **Local Name:** Mandei
                                               (M)
Medicinal Use: Dysentery (P)
Preparation and Treatment: Beniseed (*Sesamum indicum*) is pounded with young leaves of *Manniophyton fulvum* and the infusion drunk.
Family: *PIPERACEAE*
Botanical Name: *Peperomia pellucida* (L) Kunth **Local Name:** Danginye
                                               (M)
                                               Local Name: Poke leaf
                                               (K)
Medicinal Use (1): Abortion (H)
Preparation and Treatment (1): A macerate of the leaves with salt and water is drunk (~300 ml, two times a day).
Medicinal Use (2): Fertility in women (H)
Preparation and Treatment (2): A decoction of the *Peperomia pellucida* leaves and roots is use to make a rice dish with palm oil and cow's trotters and consumed as the drug as follows: The patient should sit on a mat on the floor with legs outstretched and the rice made into balls, three of which are consumed.
Medicinal Use: Athlete's foot (epidermophytosis) (Ke)
Preparation and Treatment: Leaves together with young stem are steamed and then macerated; the fluid obtained is then applied to the affected area as a lotion.

Family: *PIPERACEAE*
Botanical Name: *Piper umbellatum* Linn.   **Local Name:**
                                               Poponyamba (Ko)
Medicinal Use: Intestinal worms, particularly *Ascaris lumbricoides* (Ko)

Preparation and Treatment: The leaves are made into a sauce, with fresh fish and palm oil. The sauce is consumed as the first meal in the morning.

Family: *POACEAE (GRAMINEAE)*
Botanical Name: *Cymbopogon citratus* (DC) Stapf
Local Name: Lemon grass (K)
Medicinal Use: Fever (G)
Preparation and Treatment: An infusion of the leaves is drunk as a tea

Family: *POACEAE (GRAMINEAE)*
Botanical Name: *Pennisetum purpureum* (Schumach.)
Local Name: Bush sugar cane (K)
Medicinal Use (1): Asthma (H)
Preparation and Treatment (1): The inner part of the stem of the plant is chewed.
Medicinal Use (2): Sore throat (H)
Preparation and Treatment (2): The inner part of the stem of the plant is chewed.

Family: *POACEAE (GRAMINEAE)*
Botanical Name: *Pennisetum subangustrum* (Schumach.) Stapf & CE
Local Name: Gmamglamatotwa (L)
Medicinal Use: Wounds (G)
Preparation and Treatment: The juice obtained from crushed leaves is used as a disinfectant.

Family: *POACEAE (GRAMINEAE)*
Botanical Name: *Rottboellia exaltata* Linn.f.
Local Name: nGongo-Levu-hina (M)
Medicinal Use: Asthma (G)
Preparation and Treatment: A decoction of the young leaves with a few spices (*Xylopia aethiopica*) is drunk.

Family: *POACEAE (GRAMINEAE)*
Botanical Name: *Setaria chevalieri* Stapf & C.E. Hubb
Local Name: Bobo-Yamba (Ko)
Medicinal Use: Help weak and malnourished children to walk (Ko)

Preparation and Treatment: The grass is cut and boiled; the decoction obtained after boiling for at least one hour is the drug. The child is bathed as often as possible with the drug.

Family: *POACEAE (GRAMINEAE)*
Botanical Name: *Zea mays* Linn.   **Local Name:** Corn (K)
Medicinal Use (1): Strengthens the body (H)
Preparation and Treatment (1): The fruit is boiled in water and the liquid drunk. In addition, the patient should take a bath with the fruit of *Luffa aegyptiaca* ("sapo") (K) and a local black soap; the remaining soap and fruits are then buried.
Medicinal Use (2): Hiccups
Preparation and Treatment (2): The hairs of the corn are mixed with the eggs of a spider, heated, and then nut oil added; the drug is administered orally with the palm.

Family: *PORTULACACEAE*
Botanical Name: *Portulaca oleracea*   **Local Name:** Arata Yase (K)
Local Name: Tolugei (M)
Medicinal Use (1): Abortion (G)
Preparation and Treatment (1): A maceration of the crushed leaves in water and palm oil is drunk to cause abortion.
Medicinal Use (2): Hernia (G)
Preparation and Treatment (2): Juice obtained by squeezing slightly-heated leaves is drunk directly or after adding a few drops of palm oil.
Medicinal Use (3): Stops bleeding (H)
Preparation and Treatment (3): A decoction of the leaves of *Portulaca oleracea* and seven spices (*Xylopia*) is taken, two times a day.
Medicinal Use (4): To heal enlarged spleen in babies (H)
Preparation and Treatment (4): The leaves are first macerated and then slightly heated to warm them, and then they are applied to the area of the spleen.

Family: *RANUNCULACEAE*
Botanical Name: *Clematis grandifolia* DC   **Local Name:** Gbolo (L)

Medicinal Use: Common cold (G)
Preparation and Treatment: Leaves are crushed between the palms and the extract inhaled.

Family: *RHAMNACEAE*
Botanical Name: *Gouania longipetala* Hemsl.   **Local Name:** Sawai (M)
Medicinal Use: Fetal development (P)
Preparation and Treatment: A maceration of the crushed leaves is filtered and about 150 ml. of the filtrate drunk, three times a day, after three months pregnancy (not before).

Family: *RHIZOPHORACEAE*
Botanical Name: *Anisophyllea laurina* R. Br.   **Local Name:**
Monkiapul (K)
Medicinal Use: Improvement of milk quality (G)
Preparation and Treatment: An infusion of the leaves is used to wash the breast if the milk is suspected of being bad.

Family: *ROSACEAE/CHRYSOBALANACEAE*
Botanical Name: *Parinari excelsa* Sabine   **Local Name:** Rough skin plum (K)
Local Name: Ndawei (M)
Medicinal Use (1): Diarrhea and dysentery (F)
Preparation and Treatment (1): An infusion of the fruit is drunk.
Medicinal Use (2): Wounds (F)
Preparation and Treatment (2): The pounded bark is chewed and applied to the affected area.
Medicinal Use (3): Hernia (Ke)
Preparation and Treatment (3): The bark of the mature stem is pounded and mixed with white clay to form a solid paste. A portion of this is dissolved in water and filtered to obtain the drug, which is then drunk.
Medicinal Use (4): Stomach pain in men (H)
Preparation and Treatment (4): A decoction of the bark is drunk twice a day.

Family: *RUBIACEAE*
Botanical Name: *Amaralia heinsioides* Wernham   **Local   Name:** Gramatetei (M)
Medicinal Use (1): Eye trouble (P)
Preparation and Treatment (1): A decoction of the pounded leaves is used as an eye wash.
Medicinal Use (2): Cough (P)
Preparation and Treatment (2): The juice from gently heated leaves is drunk.

Family: *RUBIACEAE*
Botanical Name: *Borreria verticillata* (Linn.) GFW Mey   **Local Name:** Bengu (Ko)
Local   Name: Ta-latawuli (M)
Medicinal Use: Fungal disease of the skin (Tinea) "Facinclot" (K) (Ko/Ke)
Preparation and Treatment: The leaves of the plant are crushed, and the liquid obtained by squeezing it is the drug. This is applied to the infected skin twice a day after bathing (Ko). Macerated fresh leaves are applied as a lotion (Ke).

Family: *RUBIACEAE*
Botanical Name: *Canthium glabriflorum* Hiern.   **Local Name:** mBeli wa-waa (M)
Medicinal Use (1): Prolonged menstruation (P)
Preparation and Treatment (1): A decoction of the leaves or the juice from gently heated leaves is drunk to control prolonged menstruation.
Medicinal Use (2): Tonic (P)
Preparation and Treatment (2): A decoction of the leaves with those of *Vismia guineensis* is drunk.

Family: *RUBIACEAE*
Botanical Name: *Chassalia kolly* (Schumach.) Hepper   **Local Name:** Fen-Yamba (Ko)
Medicinal Use: Fever (Ko)
Preparation and Treatment: The leaves are boiled, and the decoction obtained is the drug. Approximately 100 ml is drunk two or three times a day. The drug could also be used for bathing the patient.

Family: *RUBIACEAE*
Botanical Name: *Craterispermum laurinum*
(Poir.) Benth.

Local Name: Alum bark (K)
Local Name: Alum tick (K)
Local Name: Kumbu (Ko)
Local Name: Nylei/ Nyele (M)

Medicinal Use (1): Gonorrhea (Ko)
Preparation and Treatment (1): The bark of the herb is removed, crushed, and boiled with lime (*Citrus aurantifolia*), and the liquid obtained is the drug; 100 ml of this decoction is drunk three times a day.
Medicinal Use (2): Malaria (H)
Preparation and Treatment (2): The bark of the tree is pounded, dried, and then boiled with the leaves of *Nauclea latifolia* and *Morinda geminata*; some lime juice is added and then drunk.
Medicinal Use (3): Sore throat and cough (Ko)
Preparation and Treatment (3): The young leaves are boiled, and the decoction is the drug. Approximately 150 ml is drunk twice a day.
Medicinal Use (4): Yellow fever (Ke) (P)
Preparation and Treatment (4): A decoction is made from the bark of the stem, with the epidermal layer first removed, and then drunk; this drug can also be used to steam the body and to bathe (Ke). A decoction of the stem (without epidermis) together with the leaves and stem of *Ageratum conyzoides* is drunk (P).

Family: *RUBIACEAE*
Botanical Name: *Morinda geminata* DC
(Figure 21)

Local Name: Broomstone (K)
Local Name: nJasui (M)

Medicinal Use (1): Fever (H)
Preparation and Treatment (1): A decoction of the leaves is used to bathe once daily and the steam used to heat the body.
Medicinal Use (2): Laxative (P)
Preparation and Treatment (2): A maceration of the roots or a decoction of the dried leaves alone or together with the roots of

Sarcocephalus esculentus, Smeathmannia pubescens, and *Cassia siamea* is drunk.

Medicinal Use (3): Malaria (G)

Preparation and Treatment (3): The filtrate obtained by pounding the leaves with the bark of Fignut (*Jatropha curcas*), without the epidermis, and lime is drunk.

Figure 21—*Morinda geminata*

Family: *RUBIACEAE*
Botanical Name: *Morinda morindoides* (Bak.)   **Local Name:**
Milne-Redh.   Kojo-ogbo (M)
                                                 Local Name:
                                                 Ojuologbo (K)

Medicinal Use: Malaria (Ke)

Preparation and Treatment: A decoction of the partially dried leaves and a few fruits of *Citrus aurantifolia* (lime) is taken orally.

Family: *RUBIACEAE*
Botanical Name: *Mussaenda erythrophylla*   **Local Name:**
Schum & Thonn.   Tumutumoe (Ko)

Medicinal Use (1): Convulsion in children (Ko)

Preparation and Treatment (1): A decoction of the plant together with *Selaginella myosurus* is drunk. Approximately 100 ml is taken twice a day. Also, the child may be bathed with the decoction.

Medicinal Use (2): To induce sleep in worried pregnant women (Ko)

Preparation and Treatment (2): The leaves are macerated in water, and two handfuls of the liquid are drunk when the woman reports restlessness.

Family: *RUBIACEAE*
Botanical Name: *Nauclea latifolia Sm.*   **Local Name:** A-maleke (T)
Local Name: Egbeshi (K)
Local Name: Igbesi (K)
Local Name: nNdundu (Ko)
Medicinal Use (1): Fever (H)

Preparation and Treatment (1): The roots are chopped and placed in an enamel bowl of water with seven dried peppers (*Capsicum frutesens*) and allowed to stand for seven days. One teaspoon of this infusion is drunk, twice a day.
Medicinal Use (2): Gonorrhea (Ma)
Preparation and Treatment (2): An infusion of the bark is drunk.
Medicinal Use (3): Healing of the clitoris after female circumcision and vaginal infection (H)
Preparation and Treatment (3): A decoction of *Nauclea latifolia* leaves is made, and the patient sits in a bowl of the drug three times a day.
Medicinal Use (4): Heart disease (H)
Preparation and Treatment (4): The fruits of the *Nauclea latifolia* are eaten regularly.
Medicinal Use (5): Laxative (Ko)
Preparation and Treatment (5): The leaves are boiled, and approximately 150 ml of the decoction is drunk twice a day.
Medicinal Use (6): Malaria (Ko)
Preparation and Treatment (6): The bark is placed in a container of water, and the filtrate obtained is the drug. Approximately 150 ml is taken three times a day.

Family: *RUBIACEAE*
Botanical Name: *Psychotria rufipilis* A. Chev.   **Local Name:** Kafei (M)
Medicinal Use: Pregnancy detection (P)
Preparation and Treatment: A decoction of the leaves is drunk; the drug causes movement of the fetus in the womb.
Family: *RUBIACEAE*

Botanical Name: *Rutidea olenotricha* Hiern.   **Local Name:** Waa (M)
Medicinal Use: Hemostatis (P)
Preparation and Treatment: The juice from crushed leaves stops bleeding of light wounds.

Family: *RUBIACEAE*
Botanical Name: *Sabicea vogelii* Benth.   **Local Name:** Namatei (M)
Medicinal Use: Prolonged menstruation (P)
Preparation and Treatment: A maceration of the crushed leaves or the juice from the fruits is drunk.

Family: *RUBIACEAE*
Botanical Name: *Sarcocephalus esculentus*   **Local Name:** Afzelius   Yumbuyambei (M)
Medicinal Use (1): Eye trouble (P)
Preparation and Treatment (1): The juice from the stem is used as an eye drop.
Medicinal Use (2): Laxative (P)
Preparation and Treatment (2): A maceration of the roots, which was allowed to stand for at least a day or exposed to sunlight, is drunk. Alternatively, a maceration of the roots of this plant together with the roots of *Smeathmannia pubescens*, *Morinda geminata*, and *Cassia alata* is drunk.

Family: *RUBIACEAE*
Botanical Name: *Uncaria africana* G Don   **Local Name:** Kooka (M)
Medicinal Use: Toothache (P)
Preparation and Treatment: A concentrated decoction of the leaves is used as a gargle.

Family: *RUTACEAE*
Botanical Name: *Citrus aurantifolia* (Christm.)   **Local Name:** Lem (K)
Swingle   **Local Name:** Lumbe-nyenye (M)
Local Name: Ma-roks (T)

Medicinal Use (1): Fever (G)
Preparation and Treatment (1): A warm infusion of the leaves is drunk.
Medicinal Use (2): Gonorrhea (Ma)
Preparation and Treatment (2): A number of lime fruits are cut in two and boiled until the juice is about 1/3 evaporated; A cupful of this decoction is drunk two to three times a day. This also causes purging of the patient.
Medicinal Use (3): Headache (Ma)
Preparation and Treatment (3): The leaves are pounded in a mortar, heated in an iron metal pot, and wrapped in a handkerchief and applied to the forehead.
Medicinal Use (4): Hernia (Ma)
Preparation and Treatment (4): The hernia is exposed to a hot infusion of the leaves, and the body is covered with a large cotton cloth to induce perspiration.
Medicinal Use (5): Vomiting (Ma)
Preparation and Treatment (5): A cold or warm infusion of the juice is drunk.

Family: *RUTACEAE*
Botanical Name: *Fagara macrophylla* Engl.   **Local Name:** Kpai Kine (Ko)
Medicinal Use: Impotence (Ko)
Preparation and Treatment: The bark of the tree is removed and pounded together with the bark of *Afzelia africana* and some ginger (*Zingibar officinale*). These are left in the sun and later made into a powder, which is the drug. One tablespoonful of this powder is added to hot water or palm wine (approximately 300 ml) and drunk twice a day.

Family: *SAPINDACEAE*
Botanical Name: *Allophylus africanus* P. Beauv. **Local Name:** Komi—gluei (M)
                                                   Local Name: Trilif (K)
Medicinal Use (1): Cold and headache (Ke)
Preparation and Treatment (1): The leaves are macerated between the palms, and the fumes produced are inhaled.
Medicinal Use (2): Fever (G)
Preparation and Treatment (2): An infusion of the leaves is drunk.

Medicinal Use (3): Tonic (G)
Preparation and Treatment (3): An infusion of the leaves is drunk.

Family: SAPINDACEAE
Botanical Name: *Allophylus gracilis* Radlk.   **Local Name**: Tri leaf (K)
Medicinal Use: Success in Life (H)
Preparation and Treatment: The leaves of *Allophylus gracilis* and *Paulinia pinata* are rubbed between the palms, placed in water, and then boiled. The decoction obtained is used to bathe regularly.

Family: *SAPINDACEAE*
Botanical Name: *Paullinia pinnata* Linn.   **Local Name:** Kamakagu (Ko)
Medicinal Use: Gonorrhea (Ko)
Preparation and Treatment: Leaves are pounded in a mortar, and the liquid obtained is the drug. A small amount of common salt (sodium chloride) may be added, and about one teaspoon of the concentrated drug is taken three times a day.

Family: *SAPOTACEAE*
Botanical Name: *Butyrospernum parkii* Kotschy **Local Name:** Donilif (K)
Local Name: Doni/ Shea butter (K)
Medicinal Use (1): Arthritis and Rheumatism (F)
Preparation and Treatment (1): The shea butter extracted from the seeds of this plant (also called *Vitellaria paradoxa*) is massaged on the affected areas.
Medicinal Use (2): Worms (G)
Preparation and Treatment (2): A decoction of the leaves and roots is drunk.

Family: *SCROPHULARIACEAE*
Botanical Name: *Scoparia dulcis* Linn.   **Local Name:** Pondo-livali (M)
Local Name: A-mer-ma-ka-pet (T)

Medicinal Use (1): Gonorrhea (Ma)

Preparation and Treatment (1): An infusion of the leaves is drunk.
Medicinal Use (2): Headache (Ma)
Preparation and Treatment (2): An infusion or tea of the leaves is drunk.
Medicinal Use (3): Ringworm (Ke)
Preparation and Treatment (3): Macerated leaves are prepared into a soft paste and applied as a lotion.

Family: *SCROPHULARIACEAE*
Botanical Name: *Striga macrantha* (Benth.)   **Local Name**: Efo-di-strong (K)
Medicinal Use (1): Love potion (F)
Preparation and Treatment: The plant is added to a leaf sauce and given to the lover to induce a deeper love. It also acts as an aphrodisiac.

Family: *SELAGINELLACEAE*
Botanical Name: *Selaginella myosurus* (Sw.) Alston   **Local Name:** Poro grass (K)
Local Name: Kafiyofo (Ko)
Local Name: Dimoi (M)
Medicinal Use (1): Convulsion in children (Ko)
Preparation and Treatment (1): The leaves are macerated in water with those of *Mussaenda erythropyhlla*, and the filtrate is the drug. Approximately 150 ml is taken/drunk once a day.
Medicinal Use (2): For boils (Ko)
Preparation and Treatment (2): The plant is pounded with the leaves of pepper (*Capsicum annum*) and made into a soft paste, which is rubbed evenly on the area of the boil.
Medicinal Use (3): Purgative (H)
Preparation and Treatment (3): A decoction of the leaves together with leaves of *Scleria boivinii* and the roots of *Nauclea latifolia* is drunk.
Medicinal Use (4): Skin disease (P)
Preparation and Treatment (4): A poultice obtained by crushing the whole plant is applied to the affected area. Cures skin disease (tinea) called Facinclot in Krio).

Family: *SMILACACEAE*

Botanical Name: *Smilax kraussiana* Meisn.   **Local Name:** Kandandayu (Ko)

Medicinal Use: Bilharzia (Ko)

Preparation and Treatment: Three or four young leaves are consumed with kola nuts (*Cola nitida*) three times a day.

Family: *SOLANACEAE*

Botanical Name: *Capsicum frutescens* Linn.   **Local Name:** Common red pepper/Small Pepeh (K)

Local Name: Ta-Gbengbe (T)

Medicinal Use (1): Deliverance of a person who is possessed by evil spirits. (H)

Preparation and Treatment (1): The epidermis of two roots of unequal length are scraped off, tied together, and placed in a white enamel bowl; the bowl is placed over a fire that is made up of coal and sulfur. The individual is transferred over the fire several times and the liquid spat on his/her face; the spirit is then rebuked.

Medicinal Use (2): Stops vomiting

Preparation and Treatment (2): A warm infusion is drunk or a few bruised pods are swallowed for vomiting, especially during a fever.

Family: *SOLANACEAE*

Botanical Name: *Datura metel* Linn. (Figure 22)   **Local Name:** Jao-gojie (M)/Korni (M)

Medicinal Use (1): Boils (P)

Preparation and Treatment (1): A poultice obtained by grinding leaves in water cures blue boils (P).

Medicinal Use (2): Inflammations (boils) (Ke)

Preparation and Treatment (2): The leaves are macerated and a few crystals of common salt (sodium chloride) added. The soft paste is then applied to the affected area ((Ke).

Medicinal Use (3): Snake repellent (Ke)

Preparation and Treatment (3): The whole plant is used as a snake repellent by placing it in its path. (Datura is a poisonous plant and can cause hallucinations if smoked or ingested.)

Figure 22—*Datura metel*

Family: *SOLANACEAE*
Botanical Name: *Lycopersicum esculentum* Mill.   **Local Name**: Kigongi (M)
Local Name: Tamatis (K)
Medicinal Use: Inflammation (boils) (Ke)
Preparation and Treatment: Leaves are macerated on a flat solid surface and animal fat (for instance, goat fat) is added. The paste is applied as a lotion.

Family: *SOLANACEAE*
Botanical Name: *Nicotiana sp.*   **Local Name:** Tabaka (K)
Medicinal Use (1): Hemostatis (G)
Preparation and Treatment (1): A poultice obtained by rubbing leaves between the palms is inserted into bleeding nostrils.
Medicinal Use (2): Headache (G)
Preparation and Treatment (2): The juice is squeezed dropwise into the nostrils.

Family: *SOLANACEAE*

Botanical Name: *Solanum nodiflorum* Jacq.   **Local Name:** Efo Odu (K)
Local Name: Bundu-Kemba (Ko)

Medicinal Use (1): Hemorrhoids (Ko)
Preparation and Treatment (1): The leaves are boiled, and the soft tissue of the leaf is the drug. This is eaten as part of a meal.
Medicinal Use (2): Sores (H)
Preparation and Treatment (2): A powder is made with the seeds and applied to the sore.
Medicinal Use (3): Threatened abortion (H)
Preparation and Treatment (3): A rice dish is prepared with the leaves using nut oil and cow's trotters and eaten.

Family: *STERCULIACEAE*
Botanical Name: *Cola caricaefolia* (G Don) K   **Local Name:** Kuwii (M)
Schum.
Medicinal Use: Toothache (Ke)
Preparation and Treatment: A decoction is made from the bark of the stem, and the lukewarm liquid is used to gargle at intervals.

Family: *STERCULIACEAE*
Botanical Name: *Cola nitida* (Vent.)   **Local Name:** Kola leaf (K)
Schott & Endl.
Medicinal Use (1): Eye problems (H)
Preparation and Treatment (1): The leaves are tied in a bundle, and the petioles placed in water for six hours; the resulting solution is the drug, one drop of which is applied three times a day to the eye
Medicinal Use (2): Aids movement of fetus in the womb (H)
Preparation and Treatment (2): A decoction of the dried leaves is drunk.

Family: *STERCULIACEAE*
Botanical Name: *Sterculia tragacantha* Lindl.   **Local Name:** Abala lif (K)
Medicinal Use: Sores (G)

Preparation and Treatment: Steam from a decoction of the leaves with the bark of *Parkia biglobosa* and leaves of *Mareya micrantha* is applied to the sore.

Family: *TILIACEAE*
Botanical Name: *Corchorus olitorus* Linn.   **Local Name:** nGenge
(M)
Medicinal Use: Asthma (P)
Preparation and Treatment: Crushed leaves, together with those of *Desmodium adscendeus,* are used to massage the patient.

Family: *TILIACEAE*
Botanical Name: *Triumfetta cordifolia* A Rich.   **Local Name:** Gbona
(Ko)
Medicinal Use: To stop vomiting (Ko)
Preparation and Treatment: Leaves are macerated with some water in a calabash and some common salt (sodium chloride) added. Approximately 200 ml is drunk as soon as vomiting starts.

Family: *ULMACEAE*
Botanical Name: *Trema guineensis*   **Local Name**: Gombei
(Schum. & Thonn.)   (M)
Medicinal Use: Cough (Ke)
Preparation and Treatment: The bark of the stem with epidermal layer peeled off is chewed and the juice swallowed.

Family: *UMBELLIFERAE (APIACEAE)*
Botanical Name: *Eryngium foetidum* (Linn.)   **Local Name:** Spirit lif
(Figure 23)   (K)
                          Local Name: Spirit Leaf
                          (K)
Medicinal Use (1): Fever (G)
Preparation and Treatment (1): An infusion of the leaves is drunk.
Medicinal Use (2): Protection against witches (G)
Preparation and Treatment (2): An infusion of the leaves is drunk.
Medicinal Use (3): Swellings (H)
Preparation and Treatment (3): The leaves are made into a poultice using palm oil and spice (*Xylopia*), and applied two times a day.

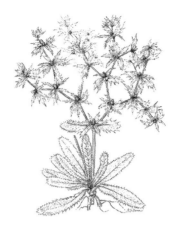

Figure 23—*Eryngium foetidum*

Family: *URTICACEAE*
Botanical Name: *Urera rigida* (Benth.) Keay   **Local Name:** Fase (Ko)
Medicinal Use: Worms (intestinal) (Ko)
Preparation and Treatment: The leaves are boiled with the leaves of lime (*Citrus aurantifolia*), and the decoction is the drug. Approximately 150 ml of the drug is taken before breakfast and also as the last drink

Family: *VERBENACEAE (LAMIACEAE)*
Botanical Name: *Clerodendron umbellatum*   **Local Name:** Bush
Poir   Bitas (K)
Medicinal Use (1): Stomach pains (G) (H)
Preparation and Treatment (1): An infusion of the leaves is drunk (G); a macerate of the leaves in water, with salt added, is drunk (H).
Medicinal Use (2): Hernia (G)
Preparation and Treatment (2): Fresh leaves are chewed.

Family: *VERBENACEAE (LAMIACEAE)*
Botanical Name: *Lantana camara* Linn.   **Local Name:** Krooman
                                              pawpaw (K)
Medicinal Use: Malaria

Preparation and Treatment : A decoction of the leaves with those of *Nauclea latiifolia*, and *Persea americana* is drunk.

Family: *VERBENACEAE (LAMIACEAE)*
Botanical Name: *Lippia adoensis* Hochst.   **Local Name:** Gambay ti-bush (K)
Medicinal Use (1): Fever (F)
Preparation and Treatment (1): The dried leaves are made into a tea and drunk.

Family: *VERBENACEAE (LAMIACEAE)*
Botanical Name: *Premna hispida* Benth.   **Local   Name:** Fono-Yamba (Ko)
Medicinal Use (1): Stops vomiting (Ko)
Preparation and Treatment (1): A decoction of the leaves is the drug. Approximately 50 ml for children and 150 ml for adults is drunk.
Medicinal Use (2): Rheumatism (Ko)
Preparation and Treatment (2): Fresh leaves are pounded in a mortar and the macerate rubbed on the joints and other sites of aches and pains.

Family: *VERBENACEAE (LAMIACEAE)*
Botanical Name: *Vitex doniana* Sweet   **Local Name:** English plum or Chuckchuck plum (K)
Medicinal Use: Sores (G)
Preparation and Treatment: Pounded leaves are tied around the sore.

Family: *ZINGIBERACEAE*
Botanical Name: *Aframomum elliotii* (Bak.) K. Schum.   **Local Name:** Ponii (M)
Medicinal Use: Measles (P)
Preparation and Treatment: A poultice obtained from the leaves and white clay is applied on the body.

Family: *ZINGIBERACEAE*
Botanical Name: *Aframomum melegueta* K Schum.   **Local Name:** Aligeta-Pepe (K)
Medicinal Use (1): Cough (G)

Preparation and Treatment (1): The seeds are chewed.

Medicinal Use (2): Stomach poison (H)

Preparation and Treatment (2): A decoction of the seeds with the leaves of other undisclosed plants is drunk.

Medicinal Use (3): Sore throat (H)

Preparation and Treatment (3): The fruit is chewed.

PART IV.
Chemical Analysis and Biological Activity of Selected Sierra Leonean Plants

Chemical, pharmacologic, and bioassay screenings have been conducted in many West African countries which share similar plant species in their flora with Sierra Leone (Oliver-Bever 1986; Soforowa 1982). Oliver-Bever grouped the plant chemicals by their actions on the cardiovascular system; those which act as stimulants/depressants on the central nervous system (CNS); those that have activities on bacteria, fungi, viruses, protozoa (antibiotic), or internal worms (anthelmintic); those with insecticidal or mulluscicidal activity; those with activity on hormone secretion with impacts on the adrenal cortex, thyroid, or sex organs; and those with hypoglycemic activity on the liver, pancreas, and kidneys to reduce blood sugar levels.

The types of compounds found in Sierra Leonean plants include phenols, quinines, flavonoids, terpenoids, alkaloids, acids, tannins, sterols, saponins, vitamins, lipids, fatty acids, essential oils, glycosides, and sterols.

In Sierra Leone, over the years, Taylor-Smith, Hanson, Roberts, Faulkner, and Koker at the chemistry department, Fourah Bay College, have conducted chemical analyses of selected Sierra Leonean plants, and some of those studies have been published: "Isolation of Anthocleistin from *Anthocleista procera* (Loganiaceae)" (Taylor-Smith 1966); "Local plants of medicinal interest, Part 4: *Habropetalum Dewei*" (Hanson 1977); "Organic constituents of *Cassia siamea*" (Roberts and Hanson, unpublished, 1977); "Epiafzelechin from the root bank of *Cassia siberiana*" (Waterman and Faulkner 1979); "Diterpenes from the stem bark of *Xylopia aethiopica*" (Faulkner, MacFoy, and Waterman 1987) and "Quinolizidine/indolizidine alkaloids from the seed of *Camoensia brevicalyx*" (Waterman and Faulkner 1981).

The author, in collaboration with Lwande et al. at the International Center for Insect Physiology and Ecology in Nairobi, Kenya, also conducted chemical analyses and biological activity studies of some African plants, much of which have been published: "9-Acridone insect antifeedants from *Teclea trichocarpa* bark" (1983); "Kaurenioc acids from *Aspilia pluriseta*" (1985); and "A new Pterocarpan from the roots of the *Tephrosia hildebrandtii*" (1987).

Examples of the compounds found in a few selected Sierra Leonean plants are presented on figures 24 through 35).

In addition, while at the biology department of Fourah Bay College, University of Sierra Leone, the author also researched and supervised several student research projects on preliminary biochemical analyses and biological activity studies of a number of Sierra Leonean plant species, for example:.

- "Antifungal activity of *Ageratum conyzoides, Ficus exasperata, Lantana camara* and *Colocasia esculentum*" (Cole 1981)
- "Insecticidal activity of *Mammea africana* and *Cucurbita pepo*" (Sama 1981)
- "Nematicidal activity of two plant species" (Bindi 1984);
- "Antifungal properties of *Ficus exasperata* and *Heliotropium indicum*" (Chaytor 1984)
- "Antibacterial properties of plants used medicinally in Gloucester village, Sierra Leone" (Cline 1985)
- "Antifungal activity of *Borreria verticillata, Erigeron floribundus, Harungana madagascariensis, Catharanthus roseus* and *Ficus exasperata*" (Lebbie 1986)
- "Antifungal and antibacterial activities of six plant species (*Aspilia africana, Carapa procera, Alchornea cordifolia, Selaginella myosurus, Lantana camara,* and *Desmodium adscendens*)" (Abanje 1986)

- "Antifungal activity of *Harungana madagascariensis, Ficus exasperata, Borreria verticilata, Selaginella myosurus, Mangifera indica,* and *Carapa procera*" (Sumana 1987)
- "Pharmacognosy of *Cassia siamea*" (Barber, 1987)
- "A review of antifungal compounds in higher plants" (MacFoy 1988)
- "In vitro antibacterial activity of six plants (*Ageratum conyzoides, Aspilia africana, Ficus exasperata, Mareya micrantha, Ocimum gratissimum,* and *Ricinus communis*) used in traditional medicine in Sierra Leone" (MacFoy and Cline 1990)

The author also showed that diterpenes from *Xylopia aethiopica* contained antimicrobial compounds (Faulkner, MacFoy, and Waterman 1987). Furthermore, he collaborated with scientists in the United Kingdom who published a paper on the antiplasmodial and antiamoebic activities of eighteen medicinal plants of Sierra Leone (Marshall et al. 2000), in addition to publishing a paper on antifungal and antibacterial activity of frog skin compounds while at American University in Washington DC, in collaboration with scientists at the National Institutes of Health in Maryland (MacFoy, Daly et al. 2005). He also collaborated with a scientist at George Washington University on preliminary anti-HIV activity of *Carapa procera* (unpublished, 2009).

In addition, Koroma and Ita (2009) have also investigated the phytochemical compounds and antimicrobial activity of three medicinal plants (*Alchornea hirtella, Morinda geminata,* and *Craterispermum laurinum*) from Sierra Leone, and more recently, Ita et al. (2010) isolated and characterized inositol from the ethanolic leaf extract of *Aspilia africana*.

Results from the above studies show that many of the plants investigated to date do show that these plants contain chemical compounds and biological activities in vitro (in test tubes), consistent with their purported use in traditional medicine, thus lending some scientific justification to their use. For example, some plants used against infection contain antibacterial activity; others used against malaria contain antimalarial/antiplasmodium activity; and plants used against fungal infection contain antifungal activity in vitro. Not surprising plants used as chewing sticks to clean the teeth where found to contain antimicrobial properties e.g. *Garcinia cola; Vernonia amygdalina* and *Zanthoxylum zanthoxyloides* (Soforowa, 1982).

It is therefore imperative that this kind of research must be sustained, as it will among other things, continue to provide new and important leads against various pharmacological targets including cancer, HIV/AIDS, Parkinson's disease, Alzheimer's, malaria, and pain, to name a few. Figures 24–35 show the structures of compounds found by scientists in selected Sierra Leonean plant species.

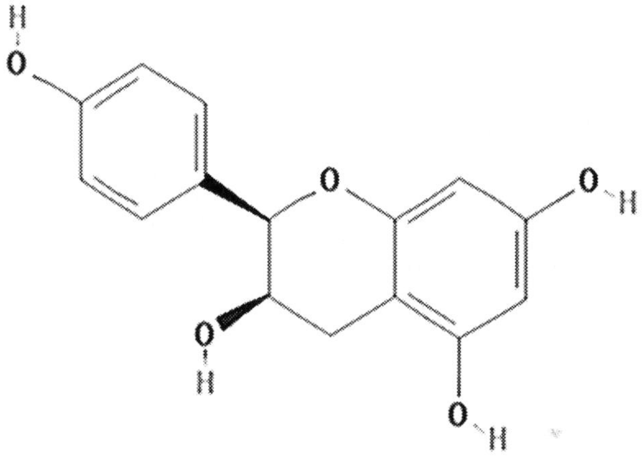

Figure 24—Epiafzelechin from the root bark of *Cassia sieberiana*

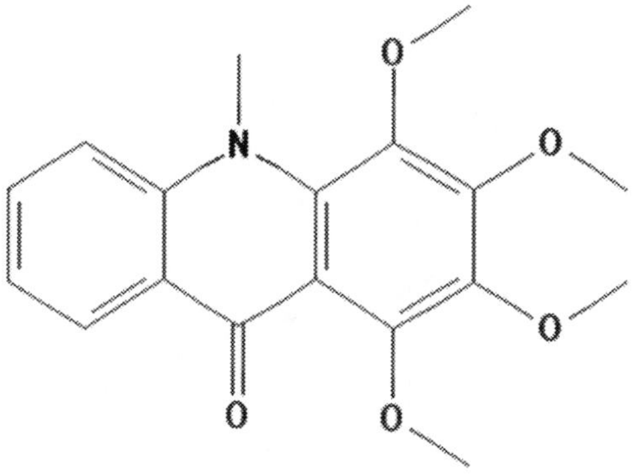

Figure 25—Melicopicine from *Teclea trichocarpa*

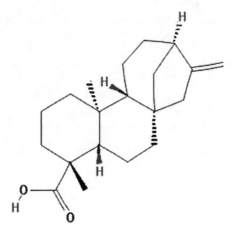

Figure 26—Kaurenoic acid from *Aspilia pluriseta*

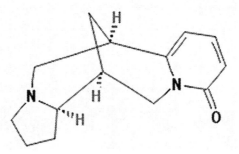

Figure 27—Camoensine from *Camoensia brevicalyx*

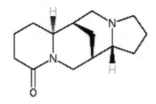

Figure 28—Camoensidine from *Camoensiabrevicalyx*

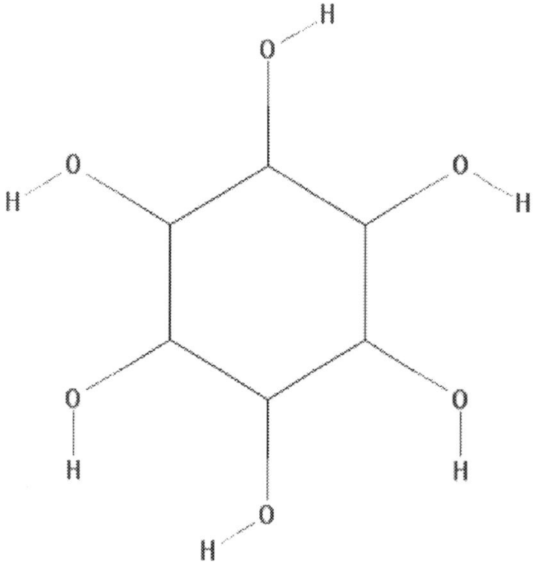

Figure 29—Inositol from *Aspilia africana*

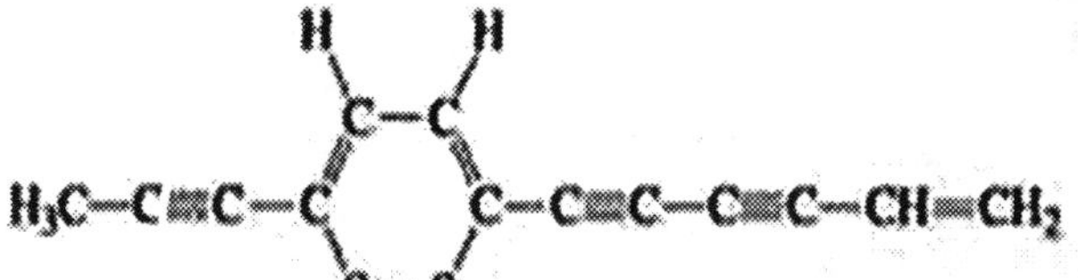

Figure 30—ThiarubineA fron *Aspilia africana*

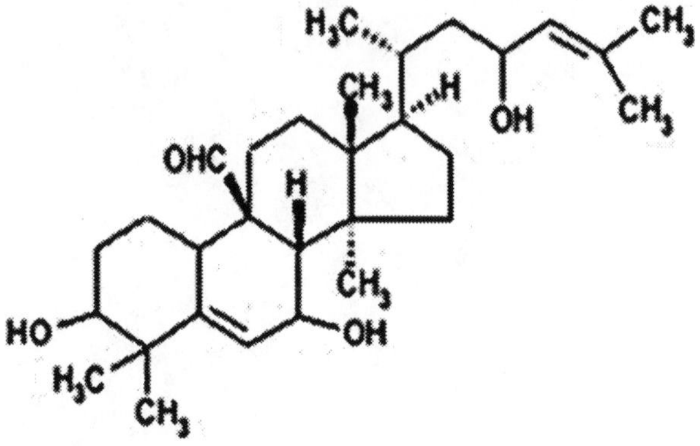

Figure 31—Momordicine from *Momordica charantia*

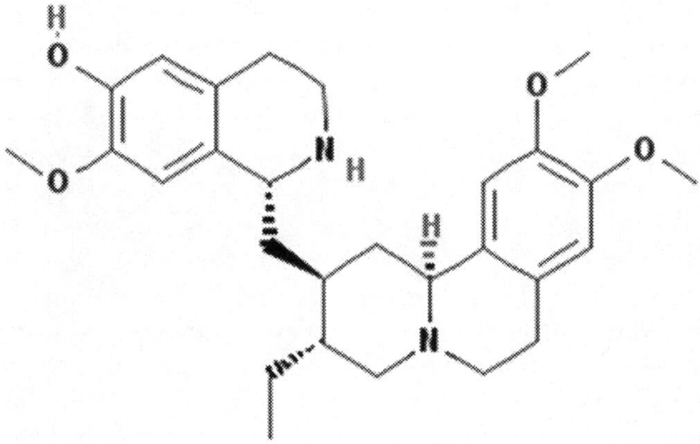

Figure 32 Cephaeline from *Borreeria verticillata*

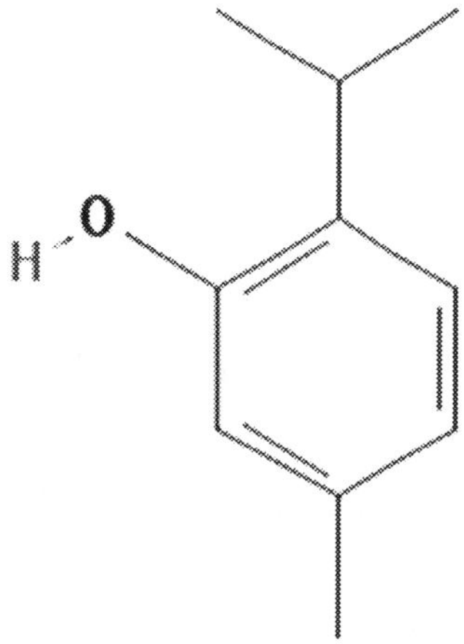

Figure 33 — Thymol from *Ocimum gratissimum*

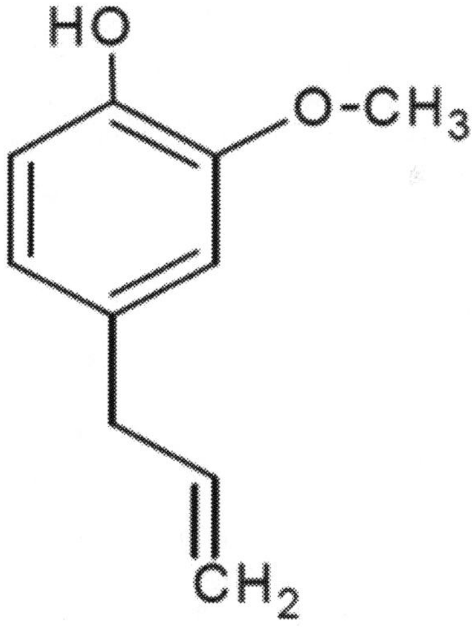

Figure 34— Eugenol from *Ocimum gratissimum*

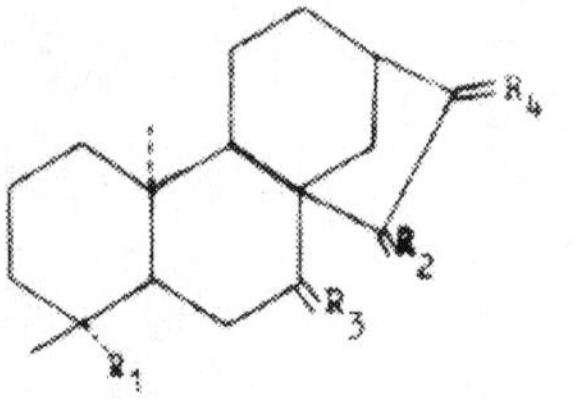

	R_1	R_2	R_3	R_4
I	COOH	β-OAc	H_2	CH_2
II	Me	H_2	H_2	α-OH; β-Me
III	COOH	H_2	H_2	CH_2
IV	COOH	OH	H_2	CH_2
V	CH_2OH	H_2	H_2	α-OH; β-Me
VI	COOH	=O	H_2	CH_2

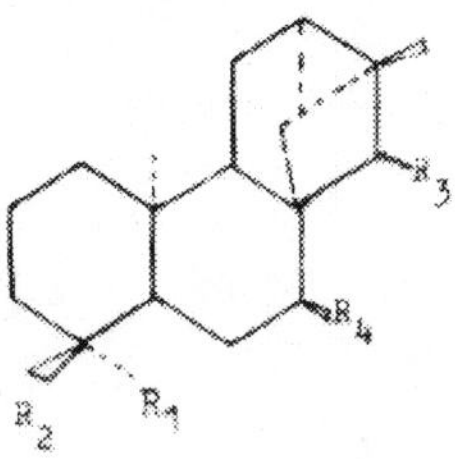

	R_1	R_2	R_3	R_4
VII	COOH	CH_3	H_2	H_2
VIII	COOH	CH_3	OH	H_2
IX	CH_3	COOH	H_2	$OCOCH_3$.

Figure 35—Some diterpenes from *Xylopia aethiopica*

Several natural product drugs derived from plants have been more recently introduced to the US market, including arteether, galantamine, nitisinone, and tiotropium; others are currently involved in late-phase clinical trials. Numerous compounds from tropical rainforest plant species with potential anticancer activity have also been identified. In addition, several compounds, mainly from edible plant species or plants used as dietary supplements that may act as chemopreventive agents, have been found. Although drug discovery from medicinal plants continues to provide an important source of new drug leads, it must be remembered that numerous challenges are encountered, including the procurement of plant materials, the selection and implementation of appropriate high-throughput screening bioassays, and the scale-up of active compounds (Balunus and Kingdom, 2005).

Hand in hand with the above, the characterization and standardization of traditional recipes for reformulation as medicines in appropriate and safe dosage is perhaps more relevant to the needs of the poor rural populations, but unfortunately, this has remained largely ignored.

Drug development programs based on ethnobotanical leads must also provide for a fair and equitable distribution of the benefits arising from this biodiversity to individual informants and local communities, especially as more and more Sierra Leonean plants have recently been reported in the popular press to be usful for various ailments e.g. bitter leaf (*Vernonia amygdalina*) for a variety of ailments such as HIV/AIDS, cancer, and diabetes; coconut oil (*Cocus nitida*) for weight loss, skin disease, ulcers and others; *Garcinia cambogia* for weight loss is in the same genus as *Garcia kola* from Sierra Leone. Sierra Leonean plants therefore have a tremendous potential to solving various health problems.

GLOSSARY

abortifacient—Induces abortion

anemia—Low red blood cell count or low hemoglobin

antidote—A substance that can be used to counteract a poison

aqueous—Of, like, or containing water

allopathic—Biological-based approach to healing; Western medicine

aphrodisiac—Increases sexual desire

asthma—An inflammation of the passages of the lungs

bilharzia (schistosomiasis)—Infection by a parasitic worm through contact with contaminated water

biodiversity—The variety of all living organisms

bioprospecting—Searching for naturally occurring chemicals from living organisms

cataplasm/poultice—Fresh parts of the plant are crushed or ground into a paste and applied to the affected area

decoction—Made by boiling crude vegetable drugs in water which is then strained or filtered to obtain the liquid drug divination—Foretelling future events or discovering hidden knowledge by occult or supernatural means

elephantiasis—A parasitic disease that is characterised by the thickening of the skin and underlying tissues, especially in the legs and genitals

ethnobotany—The study of the relationship between people and plants

ethnopharmacology -The study of the medicinal use of plants by different cultures.

febrifuge—Fever reducer

hemostasis—The stoppage of bleeding or hemorrhage

incantations—Ritual recitation of verbal charms or spells in order to produce a magic effect

inflammation—A body response characterised by redness, swelling, and heat at the site

infusion—Hot water is applied to the plant part(s) and then filtered to obtain the drug

maceration—Made by adding cold water to the drug and left standing for sometime at room temperature before straining the liquid drug

metaphysical—Reality outside the human sense perception; outside the physical world; supernatural.

morbidity—The rate of incidence of a disease

mortality—Death

poison—Capable of causing injury, illness, or death

prolotherapy—Injection of a non-active solution at the site in order to strengthen weak tissues and reduce pain

rhizome—horizontal, underground plant stem capable of producing the shoot and root stems of a new plant.

rosette—leaves radiating outwards from a short stem at soil level.

vermifuge—Expels worms

COMMON NAMES (ENGLISH)

Abrus precatorius—Bird's eye

Acacia pennata—Climbing Acacia, climbing wattle

Adansonia digitata—Baobab

Aframomum melegueta—Alligator pepper

Agelacea obliqua–Horse tamarind

Ageratum conyzoides—White head leaf; goat weed

Albizia adianthifolia—Flat-crown albizia

Alchornea cordifolia—Christmas stick

Alchornea hirtella—Forest bead-string

Allium cepa—Onion

Allophylus africanus—African false currant

Alternanthera sessilis—Sessile Joyweed, Dwarf Copperleaf

Amaranthus hybridus—Greens; spinach

Amphimas pterocarpoides—Lati; hairy tree

Anacardium occident—Cashew

Ananas comosus——Pineapple

Anchomanes difformis—Forest anchomanes

Anisophyllea laurina—Monkey apple

Annona muricata—Sour sop

Anthocleista nobilis—Cabbage tree; Cabbage palm

Anthocleista vogelii—Cabbage tree

Anthonotha macrophylla—African rosewood

Aspilia latifolia—Haemorrhage plant; wild sunflower

Aspilia pluriseta—Dwarf aspilia

Azadirachta indica—Neem tree

Bersama abyssinica—Winged Bersama

Bidens pilosa—Needle leaf; cobbler's pegs

Borreria verticallata—Shrubby false-buttonweed

Borreria verticillata,—Shrubby false-buttonweed

Bryophyllum Pinnatum—Never die; Life plant; Air plant

Butyrospermum parkii—Shea butter

Byrsocarpus coccineus—Crimson thyme

Cajanus cajan—Pigeon pea

Caloncoba brevipes—Klehn.

Calyptrochilum emarginatum—Hanging forest orchid

Capsicum frutescens—Chilies; Red pepper

Capsicum frutesens—Birdseye chili, Red pepper

Carapa procera—Crabwood

Carica papaya—Pawpaw

Catharanthus roseus—Rose periwinkle; Madagascar periwinkle

Carpodinus dulcis—Sweet pishamin; wild apricot, wild peach, rubber vine

Cassia alata—Candlebrush

Cassia occidentalis—Stinking weed; coffee senna, coffeeweed

Cassia siamea—Black-wood cassia, Bombay blackwood, cassia, iron wood, kassod tree, Siamese senna, yellow cassia

Cassia sieberiana—West Africa laburnum

Ceiba pentandra—Cotton tree

Cercestis afzelia—Cercestis

Chasmanthera dependens—Chasmanthera

Chlorophora regia—Semli

Cinnamomum zeylanicum—Cinnamum

Citrus aurantifolia—Lime

Cleistopholis patens—Apako

Clematis grandifolia—Virgins Bower; White Amenone Clematis

Clerodendrum umbellatum—Umbel Clerodendrum.

Cnestis ferruginea—Short pod, alum plant

Cocus nucifera—Coconut

Cola caricaefolia—Bumoguan Leaf

Cola nitida—Kola nut

Colocasia esculentum–Cocoyam; Taro; Eddo; Dasheen

Combretum sp.—Bush willow

Corchorus olitorus—Mallow leaves; wild jute

Costus afer—Ginger lily; 'bush-cane'

Craterispermum laurinum—Alum bark

Cucurbita pepo—Pumpkin

Cymbopogon citratus—Lemon grass

Datura metel—Devil's trumpet; hairy apple ; hairy thornapple

Desmodium adscendens—Manayupa, hardstick, amor seco, strong back, pega pega, hard man, beggar lice.

Dialium guineense—Black tumbler

Dichapetalum toxicarium—Rat's Bane

Dichrostachys glomerata—Chinese lantern; Sicklebush, Bell mimosa,

Dioscorea bulbifera—Wild Air potato; willd air yam; bitter yam

Dioscoreophyllum spp.—Guinea potato; serendipity berry

Dissotis rotundifolia-Rockrose; pink lady; Spanish Shawl

Dracaena arborea—Dragon Tree, Dracaena Tree

Dracaena sp.—Dracenia

Dracaena surculesa—Gold-dust plant" or "Gold-dust dracaena."

Elaeis guineensis—Oil palm tree

Erigeron floribundus—Bilbao fleabane; asthma weed; many-flower horse weed

Eryngium foetidum—Cilantro; Black benny

Erythrococca anomala—Bush lime;

Erythrophleum guineense—Ordeal tree; red water tree; sassy bark; sasswood tree.

Eupatorium sp.—Agueweed", "feverwort" or "sweating-plant"; bonesets, thoroughworts or snakeroots.

Euphorbia deightonii—Euphorbia

Fagara macrophylla—African satinwood

Ficus capensis-Adam's food; English fig; bush fig

Ficus exasperata—Sandpapper tree

Ficus umbellata—Broad leaf; Kennedy leaf

Garcinia cambogia—Citrin, gambooge, Brindal Berry, Gorikapuli, Malabar Tamarin

Garcinia kola—Bitter kola; false kola; male kola

Gongronema latifolium—Bush buck

Gossypium hirsutum—Cotton

Harungana madagascariensis—Blood tree

Heliotropium indicum—Cock's comb

Hibiscus esculentus—Okra

Hibiscus sabdariffa—Sorrel; Sour sour; Roselle

Hibiscus surattensis—Wild sour

Hyptis suavolens—Bush teabush

Ipomoea involucrate—Morning glory

Jatropha curcas—Physic nut

Landolphia dulcis—Sweet landolphia

Lantana camara—Lantana, largeleaf lantana, shrub verbena; red sage, yellow sage

Leonotis nepetifolia—Lion's ear; Minaret flower; Wild dagga; Throw-wort; Lion'sTail; Motherwort

Limnophyton obtusifolium—Arrow Head

Lippia adoensis—Gambian tea bush.

Lophira lanceolata—Red oak;; red ironwood

Lycopersicum esculentum—Tomato

Macaranga barteri—Macaranga

Macaranga heterophylla—Macaranga

Mammea africana—African apple, African apricot, African mammey apple, Bastard mahogany

Mammea americana—Mammy-apple

Mangifera indica—Mango

Manihot esculentum—Cassava

Manniophyton fulvum—Fulvum

Mareya micrantha—Number one

Mezoneurum benthamianum—Thorny caelapinea

Microdesmis puberula—Microdesmis

Mimosa pudica—Sensitive plant, touch-me-not plant

Momordica charantia—African Cucumber, bitter gourd, balsam pear

Morinda geminata—Brimstone bush. Indian mulberry (The Gambia, Percival).

Morinda morindoides—Brimstone tree

Musa paradisiaca—Common or horse plantain

Musa sapientum—Banana

Musanga cecropicoides—Parasolier; Umbrella tree

Mussaenda erythrophylla—shanti Blood, Red Flag Bush, Red Mussaenda, Prophet's Tears, Tropical Dogwood, Virgin Tree.

Myrainthus serratus—Myrianthus

Nauclea latifolia—Sierra Leone peach; country fig; African peach

Newbouldia laevis—Snuff–leaf; African Border Tree, Boundary Tree

Nicotiana sp.—Tobacco

Ocimum americanum-Limehairy; lemon basil or "hoary basil"

Ocimum basilicum—Sweet basil; hairy basil

Ocimum gratissimum—Tea bush, balsam, basil

Ocimum viride—Teabush

Oldfieldia Africana—African oak.

Parinari excelsa—Rough-skinned plum

Paullinia pinnata—Five finger leaf

Pennisetum purpureum—Elephant grass

Peperomia pellucida—Cow-foot-leaf

Persea Americana—Avocado Pear; Alligator pear

Phasedes lunatus—Lima bean

Piper umbellatum—Cow-foot-leaf

Portulaca oleracea—Rat's ears

Premna hispida—Headache tree

Prunus africana—African cherry

Pseudospondias microcarpa—African Grape

Psidium guajava—Guava

Raphia vinifera—Raphia and wine palm

Rauvolfia vomitoria—Swizzle-stick Tree

Ricinus communis—Castor Oil

Rottboellia exaltata—Itchgrass, Corngrass, Prickle grass

Salacia senegalensis—Stone—fruit

Sarcocephalus esculentus—African peach, Guinea peach, Sierra Leone peach

Scleria boivinii—Sword grass

Scoparia dulcis—Goatweed, scoparia-weed and sweet-broom

Sesamum indicum—Beniseed; Sesame

Setaria chevalieri—Horse grass

Sida stipulate—Stipled sida

Smilax kraussiana—West African sarsparilla

Solanum nodiflorum—White/glossy Nightshade

Sorghum bicolor—sorghum

Spondias mombin—Hog plum; fits plum

Sterculia tragacantha—African tragacanth; star chestnut;Sterculia

Striga macrantha—Witchweed or witches weed

Strychnos afzelia—Jamaica dogwood; pod mahogany . . .

Tamarindus indica—Tamarind

Tapinanthus bangwensis—Mistletoe

Teclea trichocarpa—Furry-fruited teclea

Tephrosia vogelii—Fishbean, fish-poison bean, vogel's tephrosia; fish-poison bean; wild indigo

Tetracera alnifolia—Water tree

Tetracera potatoria—Water tree

Thaumatococcus daniellii—Miracle fruit; African serendipity berry

Trema guineensis—Elm ; Pigeonwood

Triumfetta cordifolia—Burweed, cordleaf burbark

Vernonia amygdalina—Bitter leaf

Vismia guineensis—Goat's bitter leaf

Vitellaria paradoxa/Butyrospermum parkii—Shea butter tree

Vitex doniana—Black plum; West African plum.,

Xylopia aethiopica—Spice—tree

Zea mays—Maize; corn

Zingiber officinale—Ginger

BIBLIOGRAPHY

Abanje, L. (1986) Antifungal and antibacterial activities of six plant species (*Aspilia africana, Carapa procera, Alchornea cordifolia, Selaginella myosurus, Lantana camara,* and *Desmodium adscendens.* (Unpublished Manuscript).

Adevu, M. (2012) Moringa, the "Miracle Tree" is launched in Sierra Leone. http://www.umcor.org/UMCOR/Resources/News-Stories/2011/February/Moringa-the-Miracle-Tree-is-launched-in-Sierra-Leone.

"African Plant Drugs." https://tspace.library.utoronto.ca/bitstream/1807/9189/1/tc05007.pdf.

Aguwa, J. (2002). "Mende." In *Encyclopedia of World Cultures Supplement,* edited by Melvin Ember, Carol R. Ember, and Ian Skoggard, Farmington Hills:

Alie, J. (1990) A New History of Sierra Leone. Macmillan Publishers Ltd

Ayensu, E. S. (1978). *Medicinal Plants of West Africa.* Algonac: Reference Publications Inc..

Balunus, M. J., and A. D. Kingdom. (2005). "Drug Discovery from Medicinal Plants." *Life Science* 78 (5): 431- 441.

Barber. O. (1987) Pharmacognosy of *Cassia siamea"* (Unpublished Manuscript)

Barnish, G., and S. K. Samai. (1992). *Some Medicinal Plant Recipes of the Mende, Sierra Leone.* Sierra Leone: Medical Research Council Laboratory.

Baser, K. H. C., and C. A. MacFoy. (1993) "The Potential for the Industrial Utilization of Medicinal and Aromatic Plants in Sierra Leone." UNIDO Technical Report.

Bindi, A. (1984) Nematicidal activity of two plant species (Unpublished Manuscript);

CIA Report. (2013). The World Fact book. https://www.cia.gov/library/publications/the-world-factbook/geos/sl.html

Cole, M. (1981) Antifungal activity of *Ageratum conyzoides, Ficus exasperata, Lantana camara* and *Colocasia esculentum*" (Unpublished Manuscript)

Cole, N.H.A. (1968) The Vegetation of Sierra Leone. Njala University College Press. Sierra Leone.

Cole, N. H. A., and C. A. MacFoy. (1992). "Under Exploited Tropical Economic Plants in West Africa." *Journal of the Faculty of Pure and Applied Sciences* 1: 40-56.

Chaytor, T. (1984) Antifungal properties of *Ficus exasperata* and *Heliotropium indicum* (Unpublished Manuscript)

Dalziel, J. M. (1937). "Useful Plants of West Tropical Africa." London: Crown Agents for Overseas Government and Administration.

Davies, C et al (1992) Women of Sierra Leone : Traditional Voices. Partners in Adult Education Women's Commission sponsored by DVV.

Deighton, F. C. (1957). "Vernacular Botanical Vocabulary for Sierra Leone." London: Crown Agents for Overseas Government and Administration.

Dixon, P. A. F. (1981). "Medicinal Plants." In *State of Biology in Africa.*, edited by E. S. Ayensu and J. Marton-Lefevre. Washington DC, (USA). International Council of Scientific Unions

Duke, J (1998) The Green Pharmacy: The Ultimate Compendium Of Natural Remedies From The World's Foremost Authority On Healing Herbs. St. Martin's Press, New York

Duke, J. (2012). "Dr. Duke's Phytochemical and Ethnobotanical Databases." www.ars-grin.gov/duke.

EFA (2009) Environmental Foundation for Africa. Sierra Leone-Sustainable and Responsiblle Tourism. http://www.eugad.eu/wiki/index.php?title=Sierra_Leone_-_EFA_-_MDG_7,_8_-_Tiwai_Island_Wildlife_Sanctuary

Fabricant, D. S., and Norman R. Farnsworth. (2001). "The Value of Plants Used in Traditional Medicine for Drug Discovery." Supplement, *Environmental Health Perspective* 109, S1 (March).

Faulkner, D. F., C. A. MacFoy, and P. G. Waterman. (1987). "Further Diterpenes from the Stem Bark of Xylopia Aethiopica." In *Tropical Medicinal and Aromatic Plants. A Status Report*, edited by C. J. Chetsanga and C. Y. Wereko-Brobby, 338-347. Commonwealth Science Council, London, U.K.

Finch, C.S. (1990) The African Background To Medical Science. Karnak House, London, U.K.

Fyle, C.M. (2011) A Nationalist History of Sierra Leone © C. M. Fyle

Fyle, C. N., and E. D. Jones. (1980). *A Krio-English Dictionary*. Oxford University Press, Oxford, and Sierra Leone University Press, Freetown.

Hamdan, A (2009) Last date updated 17 September, 2009. Vaccines: A Religious Contention http://www.vaccineriskawareness.com/Vaccines-A-Religious-Contention-

Hanson, S. W. (1977). "Local Plants of Medicinal Interest. Part 4: *Habropetalum Dewei*." *Chemistry in Sierra Leone* 4: 38-40.

Health-SL (2012) Last date accessed: August, 2012. (http://www.health.sl/drwebsite/publish/page_69.shtml)

Herbal Medicine (2013) Last date modified: May 9th, 2013 Herbal Medicines—Best Herbal Remedies For You.by Admin Posted in Herbal Medicine http://www.greengazelles.org/page/4

Hinzen, H. (1986) *Traditional Education in Sierra Leone.* Institute of Adult Education and Extra-Mural Studies, Fourah Bay College, University of Sierra Leone, Freetown.

Ita, B., L. Koroma, and K. Kormoh. (2010). "Isolation and Characterization of Inositol from the Ethanolic Leaf Extract of *Aspilia africana*." *Journal of Chemicl and Pharaceutical Research* 2 (4): 1-6.

Kargbo, T.K. (1982) Traditional Approach to Drug Treatment in West Africa "Proceedings of the West African Pharmaceutical Federation: Third Scientific Seminar, Sierra Leone." *Nigerian Journal of Pharmacy* 13 (2) 22-26

Kokwaro, J. O. (1976). *Medicinal Plants of East Africa.* Nairobi: East Africa Literature Bureau,

Konneh, M. (2011) Sierra Leone: WHO Pledges Support for Traditional Medicine. Freetown: *Concord Times,* September 20. Freetown

Koroma, L., and B. Ita. (2009). "Phytochemical Compounds and Antimicrobial Activity of Three Medicinal Plants (*Alchornea Hirtella, Morinda Geminata,* and *Craterispermum Laurinum*) from Sierra Leone." *African Journal of Biotechnology* 8 (22): 6397-6401.

Lebbie, M. (1986) Antifungal activity of *Borreria verticillata, Erigeron floribundus, Harungana madagascariensis, Catharanthus roseus* and *Ficus exasperata*" (Unpublished)

Lebbie, A., and R. Guries. (1995). "Ethnobbotanical Value and Conservation of Saced Groves of the Kpaa Mende in Sierra Leone." *Ethnobotany* 49 (3): 297-308.

Lietava, J. (1992). "Medicinal Plants in a Middle Paleolithic Grave: Shanidar IV?" *Ethnopharmacology* 35 (3): 263-266.

Lopus M., E. Oroudjev, L. Wilson, et al. (2010) "Maytansine and Cellular Metabolites of Antibody-Maytansinoid Conjugates Strongly Suppress Microtubule Dynamics by Binding to Microtubules." *Molecular Cancer Therapy* 9 (10): 2689-2699.

Lwande, W., M. D. Bentley, C. A. MacFoy, F. N. Lugemua, A. Hasanali, and E. A. Nyandat. (1987). "A New Pterocarpan from the Roots of the *Tephrosia Hildebrandtii*." *Phytochemistry* 26: 2425-2426.

Lwande, W., T. Gebreyesus, A. Chapya, C. A. MacFoy, A. Hasanali, and M. Okech. (1983). "9-Acridone Insect Antifeedants from *Teclea Trichocarpa*." *Insect Science and Its Application* 4 (4): 393-395.

Lwande, W. A., C. A. MacFoy, M. Okech, F. Delle Monache, F. Marini Bettolo, and B. Gradgrind. 1985. "Kaurenioc Acids from *Aspilia pluriseta*." *Fitoterapia* 56 (2): 126-128.

MacFoy, C. A. 1983. *Medicinal Plants of Sierra Leone*. Freetown: Fourah Bay College, University of Sierra Leone, Sierra Leone.

MacFoy, C. A. (1985). "Medical Ethnobotany of Gloucester Village (Sierra Leone)." *Africana Research Bulletin* 14 (1 and 2): 2-14.

MacFoy C. A. (1988). "Antifungal Compounds in Higher Plants." *Journal of Chemistry in Sierra Leone* 7: 35-42

MacFoy, C. A. (1992) "Biologically Active Natural Products of Plant Origin." In *Conservation Biology,* Commonwealth Science Council, 189-194. London

MacFoy, C. A. (2004). "Ethnobotany and Sustainable Utilization of Natural Dye Plants in Sierra Leone." *Economic Botany* 58, S66-S76.

MacFoy, C. A. (2008) "Natural Dye Plants." In *Underutilized and Underexploited Horticultural Crops*, vol. 3, edited by K. V. Peter. New Delhi: NIPA., India.

MacFoy, C. A. (2010). "Comparative Ethnopharmaclogy in the Sierra Leone: American Connection." Unpublished manuscript. Microsoft Word file.

MacFoy, C. A., and E. I. Cline. (1986). "Some Medicinal Plants used in Gloucester Village (Sierra Leone) against Bacterial Infections of Man." *Journal of the Pharmaceutical Society of Sierra Leone* 1: 39-49.

MacFoy, C. A., and E. I. Cline. (1990). "In Vitro Antibacterial Activity of Three Plants Used in Traditional Medicine in Sierra Leone." *Journal of Ethnopharmacology* 28: 323-327.

MacFoy, C. A., J. Daly, et al. (2005). "Antimicrobial Activity of Frog Skin Compounds." *Zeutschrift fur Naturforschung* 60c: 932-937.

MacFoy, C. A., and A. Sama. (1983). "Medicinal Plants in Pujehun District of Sierra Leone." *Journal of Enthnopharmacology* 8: 215-223.

MacFoy C. A., and A. Sama (2012). "Insecticidal Activity of *Cucurbita Pepo* and *Mammea Africana* Plant Extracts." Unpublished article. Microsoft Word file.

Mitchel, F. (1999) Hoodoo Medicine. Summerhouse Press

Natural Standard (2013) Professional Monograph, Copyright © 2013. Nicosan: http://naturalstandard.com/index-abstract.asp?create-abstract=nicosan. asp&title=Nicosan☐

NCCAM Pub No: D456. Last Updated: September, 2011. Chronic Pain and CAM: At a Glance.

http://nccam.nih.gov/health/pain/chronic.htm

NCI (2013) Date accessed: August 12, 2013: Success Story—Taxol: http://dtp.nci.nih.gov/timeline/flash/success_stories/S2_Taxol.htm

Newton, P. (1991). "The Use of Medicinal Plants by Primates: A Missing Link?" *Trends in Ecology & Evolution* 6 (9): 297-299

Noudjou, F., et al. (2007) "Composition of *Xylopia aethiopica* (Dunal) A. Rich Essential Oils from Cameroon and Identification of a Minor Diterpene: *Ent*-13-epi Manoyl Oxide." *Biotechnol. Agron. Soc. Environ.* 11 (3): 193-199

Oliver-Bever, B. (1986) Medicinal Plants in Tropical West Africa. Cambridge University Press, Cambridge, U.K.

Pollitzer, W.S. (2005) The Gullah people and their African heritage. Athens, Georgia

Roberts, G.M.T. and S. W. Hanson. (1977). "A Preliminary Survey of Medicinal Plants of Sierra Leone." Chemistry Department, Fourah Bay College, Freetown. Unpublished manuscript. Microsoft Word file.

Sama, A. (1981) Insecticidal activity of *Mammea africana* and *Cucurbita pepo*" (Unpublished Manusript)

Soforowa, A. (1982). *Medicinal Plants and Traditional Medicine in Africa.* 2nd edition. Ibadan, Nigeria: John Wiley & Sons Ltd.

Sofowora, A. (1993a). Recent Trends in Research into African Medicinal. J. Ethnopharmacol. 38: 209-214.Referenced in A. A. Elujoba, O. M. Odeleye, and C. M. Ogunyemi. 2005. "Traditional Medicine Development for Medical and Dental Primary Health Care Delivery System in Africa." http://www.aseanbiodiversity.info/Abstract/51005777.pdf.

Stewart, K. (2003) "The African Cherry (*Prunus africana*): From Hoe Handles to the International Herb Market." *Economic Botany* 57 (4): 559-569.

Sumana, E. (1987) Antifungal activity of *Harungana madagascariensis, Ficus exasperata, Borreria verticilata, Selaginella myosurus, Mangifera indica,* and *Carapa procera*" (Unpublished Manuscript)

Taylor-Smith, R. (1966). "Isolation of Anthocleistin from *Anthocleista procera* (Loganiaceae)." *Tetrahedron* 21 (12): 3721-3725.

Taylor-Smith R. (1966). "Investigations on Plants of West Africa III. Phytochemical Studies of Some Plants of Sierra Leone." *Bull. Insst. France Afn noire A* 28: 528-541.

Taylor Smith, R. & D. E. B. Chaytor (1966) Investigations of West African Plants . III, Toxicity studies of three West African medicinal Pants. Bulletin de l'I F.A.N. T. XXVII, ser. A, no 3.

Traditional African Medicine: Spirituality (n.d.) Last date accessed: August 4, 2014: http://en.wikipedia.org/wiki/Traditional _African_Medicine

Turay, B.M.S. (1997) Medicinal Plants of Sierra Leone. A compendium. Centre for Cross- Cultural Study of Health and Healing. Alberta, Canada

Utas, M. (2009) Sexual Abuse Survivors and the Complex of Traditional Healing (G) Local Prospects in the Aftermath of an African War. Nordiska Nordiska Afrikainst itut et, UPPSALA. http://www.diva-portal.org/smash/get/diva2:217359/FULLTEXT01

Vickers, A & C. Zollman (1999) Herbal Medicine. http://www.ncbi.nlm.nih.gov/pmc/articles/PMC1116847/

Waterman, P. G., and D. F. Faulkner. (1979). "Epiafzelechin from the Root Bank of *Cassia sieberiana*." *Planta Medica* 37: 178-191.

Waterman, P. G., and D. F. Faulkner. (1982) "Quinolizidine/ Indolizidine Alkaloids from the Seed of *Camoensia Brevicalyx*." *Phytochemistry* 21 (1): 215-218.

West, K. M. (1981). "Sierra Leone: Practices of Untrained Traditional Birth Attendants and Support for TBA Training and Utilization." In *The Traditional Birth Attendant in Seven Countries: Case Studies in Utilization and Training,* edited by A. Mangay Maglacas and H. Pizurki. Geneva: WHO.

WHO (2001) Legal Status of Traditional Medicine and Complementary/Alternative Medicine: A Worldwide Review. http://apps.who.int/medicinedocs/en/d/Jh2943e/4.html

WHO (2002) WHO Traditional Medicine Strategy: 2002-2005 http://apps.who.int/medicinedocs/en/d/Js2297e/4.2.html

WHO (2013) Traditional Medicine: WHO © 2013. http://www.who.int/topics/traditional_medicine/en/

WHO-AFRO (2010) The African Health Monitor; African Traditional Medicine Day August, 31, Special Issue. http://ahm.afro.who.int/special-issue14/ahm-special-issue-14.pdf

Williams, B. E. O. (1979). "The Traditional Birth Attendant. Training and Utilization." Ministry of Health, Freetown, Sierra Leone.

Winterbottom, T. M. (1803). *An Account of the Native Africans in the Neighborhood of Sierra Leone.* C. Whittingham, London, U.K.

ABOUT THE AUTHOR

Dr. Cyrus MacFoy is a Sierra Leonean-American professor, researcher, and consultant, with more than thirty five years as a science educator in Sierra Leone, the UK, and the United States, including Fourah Bay College, University of Sierra Leone; University of London, Westminster University, Richmond College (UK); University of Maryland University College, Montgomery College, American University, Trinity University, and Johns Hopkins University (United States).

He was a researcher at the International Center for Insect Physiology and Ecology in Kenya; the United States Department of Agriculture (USDA)/NASA in Maryland; the University of Georgia; Fourah Bay College; and the University of London, with a number of publications in international journals, books, book chapters, and consultancy reports.

He has served as a consultant to various African and European countries for the United Nations, the Commonwealth Secretariat, and the World Bank.

Professor MacFoy is a graduate of Royal Holloway College, University of London, with a BS (Hons) degree in biochemistry, and of the Imperial College of Science, Technology, and Medicine, with a Masters and a Diploma of Imperial College (DIC) in environmental science and a PhD in ecological biochemistry. He is a member of the Institute of biology (MIBiol.); a chartered biologist (C.Biol.) and a Fellow of the Society of Biology (FSB).

He has attended a number of conferences, seminars, and workshops in Africa, Europe, and the United States and was visiting professor and speaker at the University of Nairobi; Howard University; USDA;

University of Georgia; University of Alabama A&M; Columbia University; the Institute of Ecosystem Studies (NY); and George Washington University. Professor MacFoy has received a number of awards and honors, including a Senior Fulbright Award; a UNIAEA award; and a Commonwealth Award, and he participated in a number of community activities in Sierra Leone, the UK, and the United States.

INDEX